16—
GP

MEANS INTERIOR ESTIMATING

Alan E. Lew

R.S. MEANS COMPANY, INC.
CONSTRUCTION CONSULTANTS & PUBLISHERS
100 Construction Plaza
P.O. Box 800
Kingston, MA 02364-0800
(617) 585-7880

© 1987

Printed in the United States of America

10   9   8   7   6   5   4

Library of Congress Card Number 87-404900

ISBN 0-87629-067-5

MEANS INTERIOR ESTIMATING

Alan E. Lew

Illustrated By
Carl W. Linde

TABLE OF CONTENTS

FOREWORD

For over 45 years, R. S. Means Company, Inc., has researched and published cost data for the building construction industry. *Means Interior Estimating* applies that valuable experience to aid the construction professional in understanding the estimating process. This book is one of a series of estimating reference books published by R. S. Means and designed to benefit all who are involved with the construction industry: the contractor, engineer, owner, developer, architect, designer, or facilities manager.

The goal of this book is to provide sound, practical methods and standards for accurate interior estimating. The entire process is demonstrated, step by step, through to the finished estimate and includes activities such as drawing takeoff, pricing, labor costing, figuring of overhead and profit, and scheduling.

The book is presented in three parts. The first part of the book, "Components of Interior Construction", is devoted to a detailed description of the different components of interior construction. A discussion of each system or component addresses the characteristics and the appropriate conditions for that component's use. A delineated takeoff procedure is provided, together with guidelines for listing and pricing materials and labor. Basic working conditions and factors are given for cost adjustments — both for increases — to allow for work over a certain height, for example — and decreases — as based on economies of scale.

The second part, "The Estimating Process", provides guidelines for estimating. This section begins with the basics, such as the types and elements of estimates. The discussion proceeds sequentially from preliminary activities through the takeoff, pricing, and estimate summary. Suggested scheduling, bidding, and project management techniques wrap up "The Estimating Process".

The third part of this book, the "Sample Estimates", uses typical drawings of interior projects as the basis for working models. Takeoff forms and procedures shown and explained in the earlier parts of the book are utilized here to complete two typical interior estimates. The costs used in the sample estimate are from the 1987 edition of Means *Interior Cost Data*. This annual Means publication — its format, use as an estimating tool, and the origin of its costs — is explained in Part Three. The sample estimates include a square foot/system budget estimate and a detailed unit price estimate.

Introduction:
BASIC
PRINCIPLES
OF ESTIMATING

The term "estimating accuracy" is a relative concept. What is the correct or accurate cost of a given construction project? Is it the total price paid to the contractor? Is it possible that another reputable contractor would perform the same work for a different cost, whether higher or lower? There is no *one* correct estimated cost for a given project. There are too many variables in construction. At best, the estimator can determine a very close approximation of what the final costs will be. The resulting accuracy of this approximation is directly affected by the amount of detail provided and the amount of time spent on the estimate.

What Is an Estimate?

Every cost estimate requires three basic components. The first is the establishment of standard *units of measure*. The second component of an estimate is the determination of the *quantity* of these units for each component, in an actual counting process: how many square yards of carpet, how many feet of partitions, number of light fixtures, etc. The third component, and perhaps the most difficult to accurately obtain, is the determination of a reasonable *cost* to purchase and install each unit.

The first element, the designation of measurement units, is the step which determines and defines the level of detail, and thus the degree of accuracy of a cost estimate. In interior construction, such units could be as all-encompassing as finished office space, or as detailed as an individual lockset. Depending on the estimator's intended use, the designation of the "unit" may describe a complete system, or it may imply only an isolated entity. The choice and detail of units also determines the time required to do an estimate.

The second component of every estimate, the determination of quantity, is more than the counting of units. In construction, this process is called the material or quantity takeoff. In order to perform this function successfully, the estimator should have a working knowledge of the materials, methods, and codes used in interior construction. An understanding of the design specifications is particularly important. This knowledge helps to assure that each quantity is correctly tabulated and that essential items are not forgotten or omitted. The estimator with a thorough knowledge of interior construction is also more likely to account for all requirements in the estimate whether or not they are called for on the plans or in the specifications. Many of the items to be quantified (counted) may not involve any material but, rather, entail labor costs only. Cleaning is an example of a labor-only item which may not be directly specified. Experience is, therefore, invaluable to ensure a complete estimate.

The third component is the determination of a reasonable cost for each unit. This aspect of the estimate is significantly responsible for variations in estimating. Rarely do two estimators arrive at exactly the same cost for a project. Total material costs in an estimate may be somewhat similar for competing contractors, however, the labor costs for installing that material will account for most variation of bids. Labor costs may vary, due to productivity as well as the pay scales in different geographical areas. The use of specialized equipment can decrease installation time and, therefore, cost. Finally, material prices, especially for finish materials and furniture, do fluctuate within the market. These cost differences occur from city to city and even from supplier to supplier in the same town. It is the experienced and well-prepared estimator who can keep track of these variations and fluctuations and use them to the best advantage when preparing accurate estimates.

This third component of the estimate, the determination of costs, can be defined in different ways by the estimator. With one approach, the estimator uses a predetermined "installed" cost which includes all the elements (material, installation, overhead, and profit) into one number expressed in dollars per unit. A variation of this approach is to use a unit cost which includes total material and installation only, adding a percent mark-up for overhead and profit to the "bottom line".

Another method is to use unit costs – in dollars – for material and for installation separately. This is done for each item, without mark-ups. These are called "bare costs". Different overhead and profit mark-ups are applied to each before the material and installation prices are added; the result would be the total "selling" price.

Types of Estimates

Interior construction estimators use four basic types of estimates. These types may be referred to by different names and may not be recognized by all as definitive. Most estimators, however, will agree that each type has its place in the estimating process. These types of estimates are as follows:

Order of Magnitude Estimate. The Order of Magnitude Estimate could be loosely described as an educated guess. It can be completed in a matter of minutes. Accuracy may be plus or minus 20%.

Square Foot and Cubic Foot Estimates. These estimates are most often useful when only the proposed size and use of an interior space is known. Square Foot and Cubic Foot estimates can be completed within an hour or two. Accuracy may be plus or minus 15%.

Systems (or Assemblies) Estimate. A Systems Estimate is best used as a budgetary tool in the planning stages of a project when some parameters have been decided. This type of estimate could require as much as one day to complete. Accuracy is expected at plus or minus 10%.

Unit Price Estimate. Working drawings and full specifications are required to complete a Unit Price Estimate. It is the most accurate of the four types but is also the most time consuming. Used primarily for bidding purposes, the accuracy of a Unit Price Estimate can be plus or minus 5%.

The figure below graphically demonstrates the relative relationship of required time versus resultant accuracy of a complete interior estimate using each of these four basic estimate types. It should be recognized that, as an estimator *and* his company gain repetitive experience on similar or identical projects, the accuracy of all four types of estimates should improve dramatically.

Order of Magnitude Estimates

The Order of Magnitude Estimate can be completed with only a minimum of information. The "units", as described earlier in this chapter, can be very general for this type and need not be well defined. For example: "The interior work for a small law firm in an urban high-rise building may cost about $50,000". This type of statement (or estimate) can be made after a few minutes of thought, based on prior experience and comparisons with similar projects from the past. While this rough figure might be appropriate for a project in one region of the country, substantial adjustments may be required for a change of geographic location. An adjustment might also be considered for cost changes over time due to inflation, changes in materials, or code changes.

Order of Magnitude costs may be developed using costs from past projects. Total project costs can be based on certain defined units: cost per attorney for law offices, cost per examining room for a medical suite, or cost per a guest room of a hotel. Experience will best help determine such costs to be used for estimating.

Square Foot and Cubic Foot Estimates

The use of Square Foot and Cubic Foot estimates is most appropriate prior to the preparation of plans or preliminary drawings, when budgetary parameters are being analyzed and established. These types of costs are broken down into different components, and then into the relationship of each component to the project as a whole, in terms of costs per square foot. This breakdown enables the designer, planner or estimator to adjust or substitute certain components according to the unique requirements of the proposed project. Most often, square foot costs include the contractor's overhead and profit.

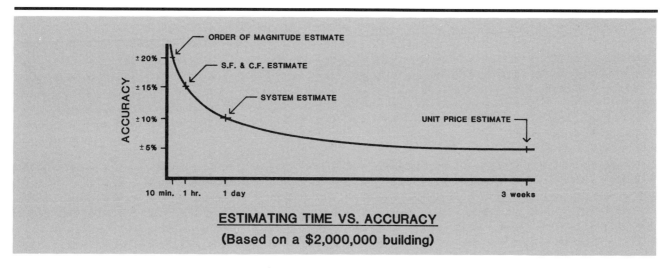

ESTIMATING TIME VS. ACCURACY
(Based on a $2,000,000 building)

Historical data for square foot costs of new building construction are plentiful. However, the best source of square foot interior costs is the estimator's own cost records for similar projects, adjusted to the parameters of the project in question. While helpful for preparing preliminary budgets, Square Foot and Cubic Foot estimates can also be useful as checks against other, more detailed estimates. While slightly more time is required than with Order of Magnitude Estimates, a greater accuracy (plus or minus 15%) is achieved due to a more specific definition of units and, therefore, the project.

Systems (or Assemblies) Estimates

One of the primary advantages of Systems (or Assemblies) Estimating is to enable alternate construction techniques to be readily compared for budgetary purposes. Rapidly rising design, product, and construction costs in recent years have made budgeting and cost effectiveness studies increasingly important in the early stages of interior projects. Never before has the estimating process had such a crucial role in the initial planning. Unit Price Estimating, because of the time and detailed information required, is not well suited as a budgetary or planning tool. A faster and more cost-effective method is needed for the planning phase of an interior project; this is the "Systems", or "Assemblies" Estimate.

The Systems method is a logical, sequential approach that reflects how a building is constructed. Twelve "Uniformat" divisions organize building construction into major components that can be used in Systems estimates. These Uniformat divisions are listed below:

Systems Estimating Divisions:

Division 1 – Foundations
Division 2 – Substructures
Division 3 – Superstructure
Division 4 – Exterior Closure
Division 5 – Roofing
Division 6 – Interior Construction
Division 7 – Conveying
Division 8 – Mechanical
Division 9 – Electrical
Division 10 – General Conditions
Division 11 – Special
Division 12 – Site Work

Each division is further broken down into systems. Division 6, which covers interior construction, includes the following major groups of fixed and moveable partitions, interior glazed openings, doors and frames, and ceilings. Each system may incorporate several different unit price components into an assemblage that is commonly used in construction.

A great advantage of the Systems Estimate is that the estimator/designer is able to substitute one system for another during design development. For example, different door systems can be compared not only by aesthetic criteria but also based on cost. The owner can anticipate and the designer can meet budget requirements before the final details and dimensions are established.

The Systems method does not require the degree of final design detail needed for a Unit Price Estimate, but estimators who use this approach must have a solid background knowledge of construction materials and methods, code requirements, design options, and budget considerations.

The Systems Estimate should not be used as a substitute for the Unit Price Estimate. While the Systems approach can be an invaluable tool in the planning stages of a project, it should be supported by Unit Price Estimating when greater accuracy is required.

Unit Price Estimates

The Unit Price Estimate is the most accurate and detailed of the four estimate types and, therefore, takes the most time to complete. Detailed working drawings and specifications must be available to the unit price estimator. All decisions regarding the project's design, materials and methods must have been made in order to complete this type of estimate. Because there are fewer variables, the estimate can be more accurate. Working drawings and specifications are used to determine the quantities of materials, equipment, and labor. Current and accurate unit costs for these items are a necessity. These costs can come from different sources. Actual quotations for similar jobs are good sources for reliable cost data. If these kinds of records are not available, prices may be determined from an up-to-date industry source book, such as Means *Interior Cost Data*.

Because of the detail involved and the need for accuracy, completion of a Unit Price Estimate entails a great deal of time and expense. For this reason, Unit Price Estimating is best suited for construction bidding. It can also be an effective method for determining certain detailed costs in a conceptual budget or during design development.

Most construction specification manuals and cost reference books, such as Means *Interior Cost Data*, compile and present unit price information into the 16 divisions of the "Masterformat" of the Construction Specifications Institute, Inc.

CSI Masterformat Divisions:

Division 1 – General Requirements
Division 2 – Site Work
Division 3 – Concrete
Division 4 – Masonry
Division 5 – Metals
Division 6 – Wood & Plastics
Division 7 – Moisture-Thermal Control
Division 8 – Doors, Windows & Glass
Division 9 – Finishes
Division 10 – Specialties
Division 11 – Equipment
Division 12 – Furnishings
Division 13 – Special Construction
Division 14 – Conveying Systems
Division 15 – Mechanical
Division 16 – Electrical

This method of organizing the various components provides a standard of uniformity that is the most widely used system in the construction industry.

A fifth type of estimate warrants mention. This is the "Scheduling Estimate", a refinement of the unit price estimate which involves the application of realistic manpower allocations. A complete unit price estimate is a prerequisite for the preparation of a scheduling estimate. The purpose of the scheduling estimate is to determine costs based upon actual working conditions. This is done using data obtained from the unit price estimate. A "human factor" can be applied. For example, if a task requires 7.5 hours to complete, based on unit price data, a tradesman will most likely take 8 hours to complete the work, and will, in any event, be paid for 8 hours work. Costs can be adjusted accordingly. A thorough discussion of scheduling and scheduling estimating is beyond the scope of this book. This subject is covered in more detail in *Means Scheduling Manual*, 2nd edition, by F. William Horsley.

Each of the 16 divisions of the Construction Specifications Institute (C.S.I.) Masterformat might be used in an interior estimate. The following chapters will review each division and highlight procedures and materials usually associated with interior estimating and construction, beginning with Division 1 – General Requirements.

Part 1
COMPONENTS OF INTERIOR CONSTRUCTION

Chapter 1
GENERAL REQUIREMENTS

When estimating by CSI division, the estimator must be careful to include all items which, while not attributable to one trade or to the physical construction of the building, are nevertheless required to successfully complete the project. These items are included in Division 1 – General Requirements. Often referred to as the "General Conditions" or "Project Overhead", they are usually set forth in the first part of the specifications. Some requirements may not be directly specified even though they are required to perform the work. Standardized sets of General Conditions have been developed by various segments of the construction industry, such as those by the American Institute of Architects, the Consulting Engineers Council/U.S., National Society of Professional Engineers, and others. These standardized documents usually include:

- Definitions
- Contract document descriptions
- Contractor's rights and responsibilities
- Architect-engineer's authority and responsibilities
- Owner's rights and responsibilities
- Variation from contract provisions
- Payment requirements and restrictions
- Requirements for the performance of the work
- Insurance and bond requirements
- Job conditions and operation

Since these documents are generic, additions, deletions and modifications unique to specific projects are usually included in the Supplementary General Conditions.

Estimated costs for Division 1 are often recorded on a standardized form or checklist. Pre-printed forms or checklists are helpful to ensure that all requirements are included and priced. Many of the costs are dependent upon work in other divisions, or on the total cost and/or duration of the job. Project overhead costs should be determined throughout the estimating process and finalized when all other divisions have been estimated and a preliminary schedule established.

The following are brief discussions of various items that may be included as project overhead. Depending upon project size and type, costs for these items may very easily represent a significant portion of a project's total costs (10% to 30% or more). The goal is to develop an approach to the estimate which will assure that all project requirements are included.

Personnel

Job site personnel may be included as either project overhead or office overhead, often depending upon the size of the project and the contractor's accounting methods. For example, if a project is large enough to require a full-time superintendent, then all costs for that person (time related) may be considered project overhead. If the superintendent is responsible for a number of smaller jobs, the expense may be either included in office overhead, or proportioned for each job.

Depending upon the size of the project, a carpenter and/or a laborer may be assigned to the job on a full-time basis for miscellaneous work. In such cases, workers are directly responsible to the job superintendent for various tasks; the costs of this work would not be attributable to any specific division and would most appropriately be included as project overhead.

Temporary Services

Required temporary services may or may not be included in the specifications. A typical statement in the specifications is: "Contractor shall supply all material, labor, equipment, tools, utilities, and other items and services required for the proper and timely performance of the work and completion of the project." As far as the owner and designer are concerned, such a statement eliminates a great deal of ambiguity. To the estimator, this means many items that must be estimated. Temporary utilities, such as heat, light, power, water, and fire protection are major considerations. The estimator must not only account for anticipated monthly (or time related) costs, but should also be sure that installation and removal costs are included, whether by the appropriate subcontractor or the general contractor.

If no room exists inside the building, an office trailer and/or storage trailers or containers may be required. Even if these items are owned by the contractor, costs should still be allocated to the job as part of project overhead. Telephone, utility, and temporary toilet facilities are other costs that must be included when required.

Depending upon the location and type of project, some security services may be required. In addition to security personnel or guard dogs, fences, gates, special lighting, and alarms may also be needed.

Temporary Construction

Temporary construction may also involve many items which are not specified in the construction documents. Temporary partitions, doors, fences, and barricades may be required to delineate or isolate portions of the building or site. In addition to these items, certain work may also be necessary solely for the protection of workers. Depending upon the project size, an OSHA representative may visit the site to assure that all safety precautions are being observed.

Workers will almost always use a new, permanent elevator for access throughout the building. While this use is almost always restricted, precautionary measures must be taken to protect the doors and cabs. Invariably, some damage occurs. Protection of any and all finished surfaces throughout the course of the project must be priced and included in the estimate. Types of temporary services and construction, with costs, from Means *Interior Cost Data*, 1987, are shown in Figure 1.1.

4

		1.1 Overhead	CREW	DAILY OUTPUT	MAN-HOURS	UNIT	BARE COSTS MAT.	LABOR	EQUIP.	TOTAL	TOTAL INCL O&P	
460	0010	SCHEDULING Critical path, as % of architectural fee, minimum				%					2%	460
	0100	Maximum				"					4%	
	0300	Computer-update, micro, no plots, minimum				Ea.					160	
	0400	Including plots, maximum				"					2,100	
	0600	Rule of thumb, CPM scheduling, small job				Job					.10%	
	0650	Large job									.05%	
	0700	Cost control, small job									.15%	
	0750	Large job									.04%	
480	0010	SMALL TOOLS As % of contractor's work, minimum ④				Total					.50%	480
	0100	Maximum				"					2%	
540	0010	TARPAULINS Cotton duck, 10 oz. to 13.13 oz. per S.Y., minimum				S.F.	.26			.26	.29	540
	0050	Maximum					.43			.43	.47	
	0100	Polyvinyl coated nylon, 14 oz. to 18 oz., minimum					.35			.35	.39	
	0150	Maximum					.43			.43	.47	
	0200	Reinforced polyethylene 3 mils thick, white					.05			.05	.06	
	0300	4 mils thick, white, clear or black					.06			.06	.07	
	0400	5.5 mils thick, clear					.07			.07	.08	
	0500	White, fire retardant					.12			.12	.13	
	0600	7.5 mils, oil resistant, fire retardant					.13			.13	.14	
	0700	8.5 mils, black					.16			.16	.18	
	0710	Woven polyethylene, 6 mils thick					.28			.28	.31	
	0740	Mylar polyester, non-reinforced, 7 mils thick					.84			.84	.92	
560	0010	TAXES Sales tax, State, County & City, average ⑧				%	4.25%					560
	0050	Maximum					7.50%					
	0200	Social Security, on first $40,000 of wages						7.15%				
	0210											
	0300	Unemployment, MA, combined Federal and State, minimum				%		2%				
	0350	Average						5.50%				
	0400	Maximum ⑤						6.20%				
	0410											
580	0010	TEMPORARY CONSTRUCTION, See division 1.5-150										580
	0300	Barricades, 5' high, 3 rail @ 2" x 8", fixed	2 Carp	30	.533	L.F.	8.80	10.95		19.75	26	
	0350	Movable	"	20	.800	"	9.50	16.45		25.95	35	
	0500	Stock units, 6' high, 8' wide, plain, buy				Ea.	353			353	390	
	0550	With reflective tape, buy				"	450			450	495	
	0600	Break-a-way 3" PVC pipe barricade										
	0610	with 3 ea. 1' x 4' reflectorized panels, buy				Ea.	250			250	275	
	1500	Plywood with steel legs, 32" wide				"	46			46	51	
	3300	Heat, incl. fuel and operation, per week, 12 hrs. per day	1 Skwk	8.75	.914	CSF Flr	14.50	19		33.50	44	
	3350	24 hrs. per day	"	4.50	1.780		19.35	37		56.35	76	
	3500	Lighting, incl. service lamps, wiring & outlets, minimum	1 Elec	34	.235		1.80	5.35		7.15	9.70	
	3550	Maximum	"	17	.471		4.15	10.65		14.80	20	
	3700	Power for temporary lighting only, per month, minimum/month								.85	.88	
	3750	Maximum/month								2.21	2.55	
	3900	Power for job duration incl. elevator, etc., minimum								42	49	
	3950	Maximum								86	100	
	4200	Office trailer, furnished, no hookups, 20' x 8', buy	2 Skwk	1	16	Ea.	3,400	335		3,735	4,225	
	4250	Rent per month	"			"				118	130	
	4700	Ramp, ¾" plywood on 2" x 6" joists, 16" O.C.	2 Carp	300	.053	S.F.	1.30	1.10		2.40	3.04	
	4750	On 2" x 10" joists, 16" O.C.	"	275	.058	"	1.95	1.20		3.15	3.90	
	5270	Stair tread protection, 2" x 12" planks, 1 use	1 Carp	75	.107	Tread	.95	2.19		3.14	4.26	
	5290	Exterior plywood, ½" thick, 1 use		65	.123		.44	2.53		2.97	4.20	
	5300	¾" thick, 1 use		60	.133		.80	2.74		3.54	4.90	
	5500	Storage vans, trailer mounted, 16' x 8', buy	2 Skwk	1.80	8.890	Ea.	2,265	185		2,450	2,775	
	5520	Rent per month				"	77			77	85	
620	0010	WATCHMAN Service, monthly basis, uniformed man, minimum				Hr.					6.10	620
	0100	Maximum				"					11.35	
680	0010	WINTER PROTECTION Reinforced plastic on wood										680
	0100	framing to close openings	2 Clab	750	.021	S.F.	.24	.34		.58	.77	

4

For expanded coverage of these items see *Means Building Construction Cost Data 1987*

Figure 1.1

Job Cleanup

An amount should always be carried in the estimate for cleanup of the construction area, both during the construction process and upon completion. The cleanup can be broken down into three basic categories, and these can be estimated separately:

- Continuous (daily or otherwise) cleaning of the project area
- Rubbish handling and removal
- Final cleanup

Costs for continuous cleaning can be included as an allowance, or estimated by required man-hours. Rubbish handling should include barrels, a trash chute if necessary, dumpster rental, and disposal fees. These fees vary depending upon the project, and a permit may also be required. Costs for final cleanup should be based upon past projects and may include subcontract costs for items such as the cleaning of windows and waxing of floors. Included in the costs for final cleanup may be an allowance for repair of minor damage to finished work.

Bonds

Bonding requirements for a project will be specified in Division 1 – General Requirements, and will be included in the construction contract. Various types of bonds may be required. Listed below are a few common types:

Bid Bond. A form of bid security executed by the bidder or principle and by a surety (bonding company) to guarantee that the bidder will enter into a contract within a specified time and furnish any required Performance of Labor and Material Payment bonds.

Completion Bond. Also known as "Construction" or "Contract" bond. The guarantee by a surety to the owner that the contractor will pay for all labor and materials used in the performance of the contract as per the construction documents. The claimants under the bond are those having direct contracts with the contractor or any subcontractor.

Performance Bond. (1) A guarantee that a contractor will perform a job according to the terms of the contracts. (2) A bond of the contractor in which a surety guarantees to the owner that the work will be performed in accordance with the contract documents. Except where prohibited by statute, the performance bond is frequently combined with the labor and material payment bond.

Surety Bond. A legal instrument under which one party agrees to answer to another party for the debt, default or failure to perform of a third party.

Miscellaneous General Requirements

Many other items must be taken into account when costs are being determined for project overhead. Among the major considerations are:

Scaffolding or Rolling Platforms. It is important to determine who is responsible for rental, erection, and dismantling of scaffolding, since it will often be used by more than one trade. If a subcontractor is responsible, it may be necessary to leave the scaffolding in place long enough for use by other trades. Different types of scaffolding are illustrated in Figure 1.2.

Small Tools. An allowance, based on past experience, should be carried for small tools. This allowance should cover hand tools as well as small power tools for use by workers on the interior contractor's payroll. Small tools have a habit of "walking" and a certain amount of replacement is necessary.

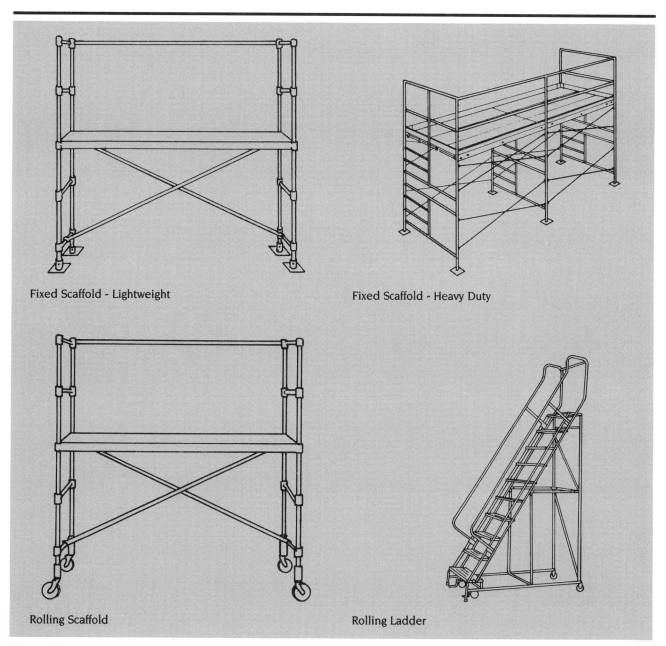

Fixed Scaffold - Lightweight

Fixed Scaffold - Heavy Duty

Rolling Scaffold

Rolling Ladder

Figure 1.2

Permits. Various types of permits may be required depending upon local codes and regulations. Both the necessity of the permit and the responsibility for acquiring it must be determined. If the work is being done in an unfamiliar location, local building officials should be consulted regarding unusual or unknown requirements.

Insurance. Insurance coverage for each project and locality – above and beyond normal, required operating insurance – should be reviewed to assure that coverage is adequate. The contract documents will often specify certain required policy limits. The need for specific policies or riders should be anticipated.

Other items commonly included in project overhead are: photographs, models, job signs, costs for sample panels and materials for owner/architect approval, and an allowance for replacement of broken glass. For some materials, such as imported goods or custom fabricated items, both shipping costs and off site storage fees can be expected. An allowance should be included for anticipated costs pertaining to punch list items. These costs are likely to be based on past experience.

Some project overhead costs can be calculated at the beginning of the estimate. Others will be included as the estimating process proceeds. Still other costs are estimated last since they are dependent upon the total cost and duration of the project. Because many of the overhead items are not directly specified, the estimator must draw from experience and visualize the construction process to assure that all requirements are met. It is not important when or where these items are included, but that they are included. One contractor may list certain costs as project overhead, while another contractor would allocate the same costs (and responsibility) to a subcontractor. Either way, the costs are recorded in the estimate.

Chapter 2

SITE WORK AND DEMOLITION

Site work is not often included in interior projects. However, entrances with paved areas as exterior/interior transition surfaces might be considered part of the interiors contract. Examples are shown in Figure 2.1, from Means *Interior Cost Data*, 1987. Mezzanine additions may require additional footings that might entail the removal of concrete slabs, excavation, and backfill. Access for such work is often severely restricted and must be carefully evaluated so that appropriate costs are included in the estimate.

However small, the amount of site work involved should not be of less importance during the quantity takeoff. Because most site work requires special tools or heavy equipment (shown in Figure 2.2), its role in interior work requires special attention. The estimator must determine the cost effectiveness of such equipment. Minimum charges may be in effect if the quantities of work are small. Not only must the estimator determine exact quantities involved, but he must also employ careful judgement based on experience to anticipate consequences.

Demolition may have to be carefully broken down into individual components that are not directly included in the plans and specifications. Unless the estimator has had extensive experience in demolition, or unless a local subcontractor can provide a bid, each item to be demolished should be listed separately. When performing the quantity takeoff, requirements for handling, hauling, and dumping the debris must also be included. Figure 2.3, from Means *Interior Cost Data*, 1987, shows costs for the demolition of certain types of walls and partitions. In addition, costs for rubbish handling are shown. These costs must be added *separately* to the demolition costs to be sure that the estimate is complete.

SITE WORK	12.7-100	Walkways

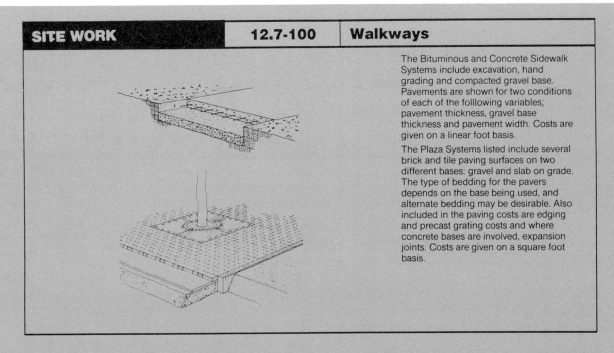

The Bituminous and Concrete Sidewalk Systems include excavation, hand grading and compacted gravel base. Pavements are shown for two conditions of each of the folllowing variables; pavement thickness, gravel base thickness and pavement width. Costs are given on a linear foot basis.

The Plaza Systems listed include several brick and tile paving surfaces on two different bases: gravel and slab on grade. The type of bedding for the pavers depends on the base being used, and alternate bedding may be desirable. Also included in the paving costs are edging and precast grating costs and where concrete bases are involved, expansion joints. Costs are given on a square foot basis.

12.7-120	Bituminous Sidewalks	COST PER L.F.		
		MAT.	INST.	TOTAL
1580	Bituminous sidewalk, 1″ thick paving, 4″ gravel base, 3′ width	.69	1.53	2.22
1600	4′ width	.92	2.01	2.93
1640	6″ gravel base, 3′ width	.77	1.68	2.45
1660	4′ width	1.01	2.21	3.22
2120	2″ thick paving, 4″ gravel base, 3′ width	1.24	1.93	3.17
2140	4′ width	1.66	2.50	4.16
2180	6″ gravel base, 3′ width	1.32	2.07	3.39
2200	4′ width	1.75	2.71	4.46

12.7-140	Concrete Sidewalks	COST PER L.F.		
		MAT.	INST.	TOTAL
1580	Concrete sidewalk, 4″ thick, 4″ gravel base, 3′ wide	2.70	2.88	5.58
1600	4′ wide	3.59	3.75	7.34
1640	6″ gravel base, 3′ wide	2.78	3.03	5.81
1660	4′ wide	3.68	3.95	7.63
2120	6″ thick concrete, 4″ gravel base, 3′ wide	3.84	2.97	6.81
2140	4′ wide	5.05	3.83	8.88
2180	6″ gravel base, 3′ wide	3.92	3.12	7.04
2200	4′ wide	5.15	4.29	9.44

12.7-283	Brick & Tile Plazas	COST PER S.F.		
		MAT.	INST.	TOTAL
2050	Brick pavers, 4″ x 8″ x 1-½″, gravel base, stone dust bedding	2.56	2.48	5.04
2100	Slab on grade, asphalt bedding	3.63	3.02	6.65
3550	Concrete paving stone, 4″ x 8″ x 2-½″, gravel base, sand bedding	3.06	2.46	5.52
3600	Slab on grade, asphalt bedding	4.08	2.94	ι7.02
4050	Concrete patio blocks, 8″ x 16″ x 2″, gravel base, sand bedding	1.69	2.29	3.98
4100	Slab on grade, asphalt bedding	2.71	2.77	5.48
6050	Granite pavers, 3-½″ x 3-½″ x 3-½″, gravel base, sand bedding	5.60	5.50	11.10
6100	Slab on grade, mortar bedding	6.50	6.05	12.55

Figure 2.1

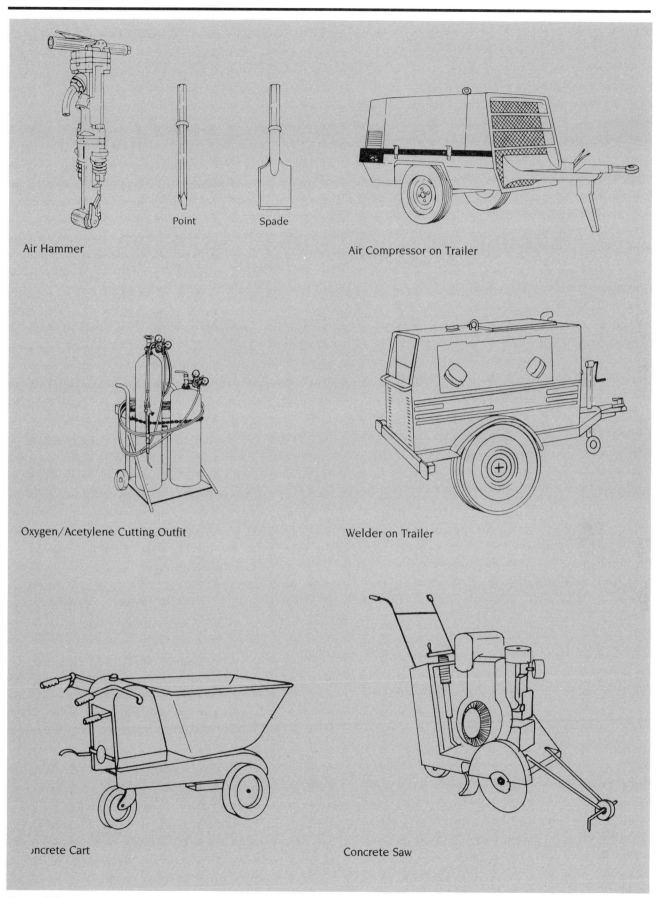

Air Hammer

Point

Spade

Air Compressor on Trailer

Oxygen/Acetylene Cutting Outfit

Welder on Trailer

Concrete Cart

Concrete Saw

Figure 2.2

		2.2 Building Demolition	CREW	DAILY OUTPUT	MAN-HOURS	UNIT	BARE COSTS				TOTAL INCL O&P	
							MAT.	LABOR	EQUIP.	TOTAL		
640	5240	Shingles	1 Clab	250	.032	S.F.		.52		.52	.76	640
	5260	Textured plywood	"	500	.016	"		.26		.26	.38	
650	0010	RUBBISH HANDLING The following are to be added to the										650
	0020	demolition prices										
	0400	Chute, circular, prefabricated steel, 18″ diameter	B-1	40	.600	L.F.	8.75	10.05		18.80	24	
	0440	30″ diameter	"	30	.800	"	17	13.40		30.40	38	
	0600	Dumpster, (debris box container), 5 C.Y., rent per week				Ea.					85	
	0700	10 C.Y. capacity									110	
	0800	30 C.Y. capacity									150	
	0840	40 C.Y. capacity									185	
	1000	Dust partition, 6 mil polyethylene, 4′ x 8′ panels, 1″ x 3″ frame	2 Carp	2,000	.008	S.F.	.15	.16		.31	.41	
	1080	2″ x 4″ frame	"	2,000	.008	"	.25	.16		.41	.52	
	2000	Load, haul to chute & dumping into chute, 50′ haul	2 Clab	21.50	.744	C.Y.		12		12	17.60	
	2040	100′ haul		16.50	.970			15.60		15.60	23	
	2080	Over 100′ haul, add per 100 L.F.		35.50	.451			7.25		7.25	10.65	
	2120	In elevators, per 10 floors, add		140	.114			1.84		1.84	2.70	
	3000	Loading & trucking, including 2 mile haul, chute loaded	B-16	32	1			16.80	8.70	25.50	34	
	3040	Hand loaded, 50′ haul	2 Clab	30	.533			8.60		8.60	12.60	
	3080	Machine loaded	B-6	60	.400			6.90	2.70	9.60	13.10	
	3120	Wheeled 50′ and ramp dump loaded	2 Clab	24	.667			10.75		10.75	15.75	
	5000	Haul, per mile, up to 8 C.Y. truck	B-34B	1,165	.007			.12	.24	.36	.43	
	5100	Over 8 C.Y. truck	"	1,550	.005			.09	.18	.27	.32	
660	0010	SAW CUTTING Asphalt over 1000 L.F., 3″ deep	B-89	640	.025	L.F.	.20	.46	.04	.70	.95	660
	0020	Each additional inch of depth		1,035	.015		.05	.29	.03	.37	.51	
	0400	Concrete slabs, mesh reinforcing, per inch of depth		795	.020		.24	.37	.03	.64	.85	
	0420	Rod reinforcing, per inch of depth		455	.035		.32	.65	.06	1.03	1.37	
	0800	Concrete walls, plain, per inch of depth		170	.094		.22	1.74	.16	2.12	2.98	
	0820	Rod reinforcing, per inch of depth		135	.119		.32	2.19	.21	2.72	3.80	
	1200	Masonry walls, brick, per inch of depth		140	.114		.22	2.11	.20	2.53	3.56	
	1220	Block walls, solid, per inch of depth		170	.094		.22	1.74	.16	2.12	2.98	
	5000	Wood sheathing to 1″ thick, on walls	1 Carp	200	.040			.82		.82	1.21	
	5020	On roof	"	250	.032			.66		.66	.97	
	9950	See also Div. 2.1-200 core drilling										
670	0010	TORCH CUTTING Steel, 1″ thick plate	A-1	95	.084	L.F.		1.36	.42	1.78	2.46	670
	0040	1″ diameter bar	"	210	.038	Ea.		.61	.19	.80	1.11	
	1000	Oxygen lance cutting, reinforced concrete walls										
	1040	12″ to 16″ thick walls	1 Clab	10	.800	L.F.		12.90		12.90	18.90	
	1080	24″ thick walls	"	6	1.330	"		21		21	32	
680	0010	WALLS AND PARTITIONS DEMOLITION										680
	0020											
	0100	Brick, 4″ to 12″ thick	B-9	220	.182	C.F.		3	.67	3.67	5.15	
	0150											
	0200	Concrete block, 4″ thick	B-9	1,000	.040	S.F.		.66	.15	.81	1.13	
	0280	8″ thick	"	810	.049			.81	.18	.99	1.40	
	1000	Drywall, nailed	1 Clab	1,000	.008			.13		.13	.19	
	1020	Glued & nailed		900	.009			.14		.14	.21	
	1500	Fiberboard, nailed		900	.009			.14		.14	.21	
	1520	Glued & nailed		800	.010			.16		.16	.24	
	2000	Movable walls, metal, 5′ high		300	.027			.43		.43	.63	
	2020	8′ high		400	.020			.32		.32	.47	
	2200	Metal or wood studs, finish 2 sides, fiberboard	B-1	520	.046			.77		.77	1.14	
	2250	Lath and plaster		260	.092			1.55		1.55	2.27	
	2300	Plasterboard (drywall)		520	.046			.77		.77	1.14	
	2350	Plywood		450	.053			.89		.89	1.31	
	2900	Plasterboard (drywall), one side		2,000	.012			.20		.20	.30	
	2950											
	3000	Plaster, lime and horsehair, on wood lath	1 Clab	400	.020	S.F.		.32		.32	.47	
	3020	On metal lath	"	335	.024	"		.38		.38	.56	

For expanded coverage of these items see *Means Site Work Cost Data 1987*

15

Figure 2.3

Chapter 3
CONCRETE

Concrete work in interior construction may entail placing a floor slab, placing a new topping on an existing floor, constructing new column foundations, or providing equipment foundations. Figure 3.1 illustrates different types of concrete work commonly used in interior projects.

Concrete work can be estimated in two basic manners. All of the individual components – formwork, reinforcing, concrete, placement, finish – can be estimated separately. This is the most accurate, but most time-consuming method and is not always practical when the concrete work is only a small portion of the total interior project. The second method is to estimate using costs which incorporate all of the components into a system or assembly. Such costs are shown in Figure 3.2. When using this method, the estimator must be sure the system being estimated is the same as that for which costs are developed. Slight variations in design of the system can significantly affect costs.

Regardless of the scope of the work required, all concrete construction involves the following basic components:

- Formwork
- Reinforcing steel or welded wire fabric
- Concrete of a specified compressive strength
- Placement
- Specified finish if necessary

Following are descriptions of various concrete design details found in interior construction. Reinforcing, strength, placement, and finish requirements are included where applicable.

Floor slabs (on grade) should be placed on compacted granular fill, such as gravel or crushed stone, which in turn is covered by a polyethylene vapor barrier. Slabs may be reinforced or unreinforced. Reinforcing is usually provided by welded wire fabric, primarily to provide assurance that any crack that does occur will be tightly closed.

Large concrete floors are usually placed in strips which extend across the building at column lines or at 20' to 30' widths. Construction joints may be keyed or straight and may contain smooth or deformed reinforcing which, in turn, can be wrapped or greased on one side to allow horizontal movement and to control cracking.

Control joints hopefully limit cracking to designated lines and may be established by saw-cutting the partially cured concrete slab to a specified depth or by applying a preformed metal strip to create a crack line. Many specifications require a boxed out section around columns, with that area to be concreted after the slab has been placed and cured.

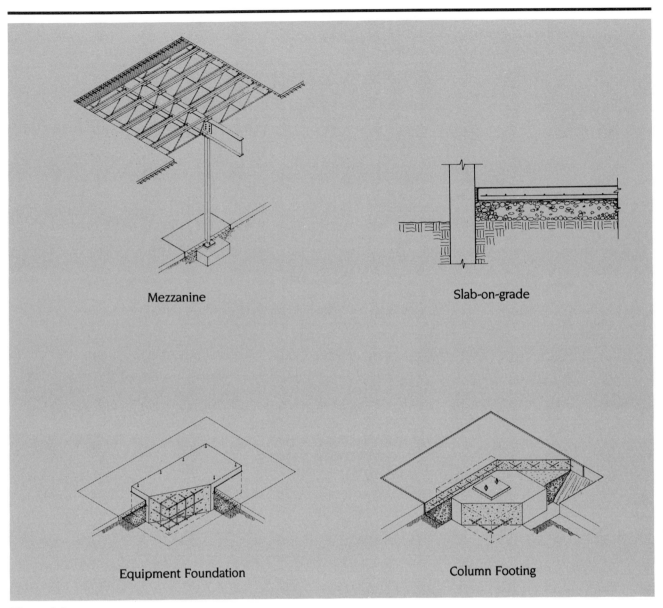

Mezzanine

Slab-on-grade

Equipment Foundation

Column Footing

Figure 3.1

3.3	Cast in Place Concrete	CREW	DAILY OUTPUT	MAN-HOURS	UNIT	MAT.	LABOR	EQUIP.	TOTAL	TOTAL INCL O&P	
080	0100	Silicon carbide, black, over 20 tons				Lb.	.85			.85	.94	080
	0120	Under 1 ton					.90			.90	.99	
	0500	Carbon black, liquid, 2 to 8 lbs. per bag of cement					2.35			2.35	2.59	
	0600	Colors, integral, 2 to 10 lb. per bag of cement, minimum					.80			.80	.88	
	0610	Average					1.30			1.30	1.43	
	0620	Maximum					3			3	3.30	
	0700	Curing compound, 200 to 400 S.F. per gallon, 55 gal. lots				Gal.	5.45			5.45	6	
	0720	5 gallon lots					6.50			6.50	7.15	
	0800	Premium grade, 450 S.F. per gallon, 55 gallon lots					12			12	13.20	
	0820	5 gallon lots					12.60			12.60	13.85	
	0900	Dustproofing compound, 200-600 S.F./gal., 55 gallon lots					6			6	6.60	
	0920	5 gallon lots					7.20			7.20	7.90	
	1000	Epoxy dustproof coating, colors, 300-400 S.F. per coat,										
	1010	or transparent, 400-600 S.F. per coat				Gal.	45			45	50	
	1100	Hardeners, metallic, 55 lb. bags, natural (grey)				Lb.	.35			.35	.39	
	1200	Colors, average					.55			.55	.61	
	1300	Non-metallic, 55 lb. bags, natural (grey), minimum					.30			.30	.33	
	1310	Maximum					.40			.40	.44	
	1320	Non-metallic, colors, mininum					.35			.35	.39	
	1340	Maximum					.45			.45	.50	
	1400	Non-metallic, non-slip, 100 lb. bags, minimum					.35			.35	.39	
	1420	Maximum					.45			.45	.50	
	1500	Solution type, 300 to 400 S.F. per gallon				Gal.	4.80			4.80	5.30	
	1510											
	1600	Sealer, hardener and dustproofer, clear, 450 S.F., minimum				Gal.	7			7	7.70	
	1620	Maximum					17			17	18.70	
	1700	Colors (300-400 S.F. per gallon)					15.75			15.75	17.35	
	1710											
	2000	Waterproofing, integral 1 lb. per bag of cement				Lb.	.75			.75	.83	
	2100	Powdered metallic, 40 lbs. per 100 S.F., minimum					.80			.80	.88	
	2120	Maximum					1.20			1.20	1.32	
	2200	Water reducing admixture, average				Gal.	7.75			7.75	8.55	
140	0010	CONCRETE IN PLACE Including forms (4 uses), reinforcing										140
	0050	steel, including finishing unless otherwise indicated										
	0150	Average for superstructure only, including finishing	C-17B	13.42	6.110	C.Y.	107	130	17.95	254.95	330	
	0600											
	0700	Columns, square, 12″ x 12″, minimum reinforcing	C-17A	4.60	17.610	C.Y.	185	375	27	587	785	
	0800	16″ x 16″, minimum reinforcing		6.25	12.960		165	275	20	460	610	
	0900	24″ x 24″, minimum reinforcing		9.08	8.920		145	190	13.80	348.80	455	
	3800	Footings, spread under 1 C.Y.	C-17B	31.82	2.580		74	55	7.60	136.60	170	
	3850	Over 5 C.Y.	C-17C	70.45	1.180		70	25	5.20	100.20	120	
	3900	Footings, strip, 18″ x 9″, plain	C-17B	34.22	2.400		63	51	7.05	121.05	150	
	3950	36″ x 12″, reinforced		49.07	1.670		69	35	4.91	108.91	135	
	4000	Foundation mat, under 10 C.Y.		32.32	2.540		115	54	7.45	176.45	215	
	4050	Over 20 C.Y.		47.37	1.730		105	37	5.10	147.10	175	
	4200	Grade walls, 8″ thick, 8′ high	C-17A	10.16	7.970		105	170	12.35	287.35	380	
	4260	12″ thick, 8′ high	″	13.50	6		125	125	9.30	259.30	335	
	4300	15″ thick, 8′ high	C-17B	20.01	4.100		85	87	12.05	184.05	235	
	4650	Ground slab, not including finish, 4″ thick	C-17C	75.28	1.100		63	23	4.87	90.87	110	
	4700	6″ thick	″	113.47	.731		60	15.50	3.23	78.73	92	
	4750	Ground slab, incl. troweled finish, not incl. forms										
	4760	or reinforcing, over 10,000 S.F., 4″ thick slab	C-8	3,520	.016	S.F.	.73	.29	.14	1.16	1.37	
	4820	6″ thick slab		3,610	.016		1.10	.28	.14	1.52	1.77	
	4840	8″ thick slab		3,275	.017		1.48	.31	.15	1.94	2.25	
	4900	12″ thick slab		2,875	.019		2.20	.35	.17	2.72	3.12	
	4950	15″ thick slab		2,560	.022		2.75	.40	.19	3.34	3.82	
	6800	Stairs, not including safety treads, free standing	C-15	120	.600	LF Nose	4.60	11.55	.40	16.55	22	
	6850	Cast on ground		180	.400	″	3.45	7.70	.26	11.41	15.40	
	7000	Stair landings, free standing		285	.253	S.F.	1.90	4.87	.17	6.94	9.45	
	7050	Cast on ground		685	.105	″	1.12	2.03	.07	3.22	4.28	

24

For expanded coverage of these items see *Means Concrete Cost Data 1987*

Figure 3.2

Expansion joints are generally used against confining walls, foundations, etc., and are commonly made from a preformed expansion material. The finish of the slabs is usually dictated by their use and varies between a screed (rough) finish (associated with two-course floors) to a steel trowel-treated finish (common for exposed concrete floors). Dropped areas or deeper slabs may be used under concentrated loads, such as masonry walls. Formed depressions to receive other floor materials, such as mud-set ceramic tile or terazzo, may also be necessary, but at an additional cost to the slab-on-grade system.

Concrete floors on slab form, or centering, is a widely used structural system because of its light weight, fast erection time and flexibility in bay sizes. Figure 3.3 illustrates typical concrete construction details.

Concrete slabs (elevated) are normally placed over steel forms attached to the joists or beams by welding or by self-tapping screws. Slabs are normally reinforced with welded-wire fabric. Figure 3.4 illustrates such a system for a mezzanine added in an existing building. Note all the components listed that must be included in a complete estimate for this type of work.

Concrete toppings may include new concrete slabs placed over existing floors, with colors, hardeners, integral toppings, or abrasive finishes. Granolithic topping 1/2" to 2" in thickness may be placed over existing slabs with the proper surface preparation. Other concrete work which may be required should be estimated separately. Such work may include locker bases, closures around pipes in pipe chases, required grouting around door sills and recessed hardware.

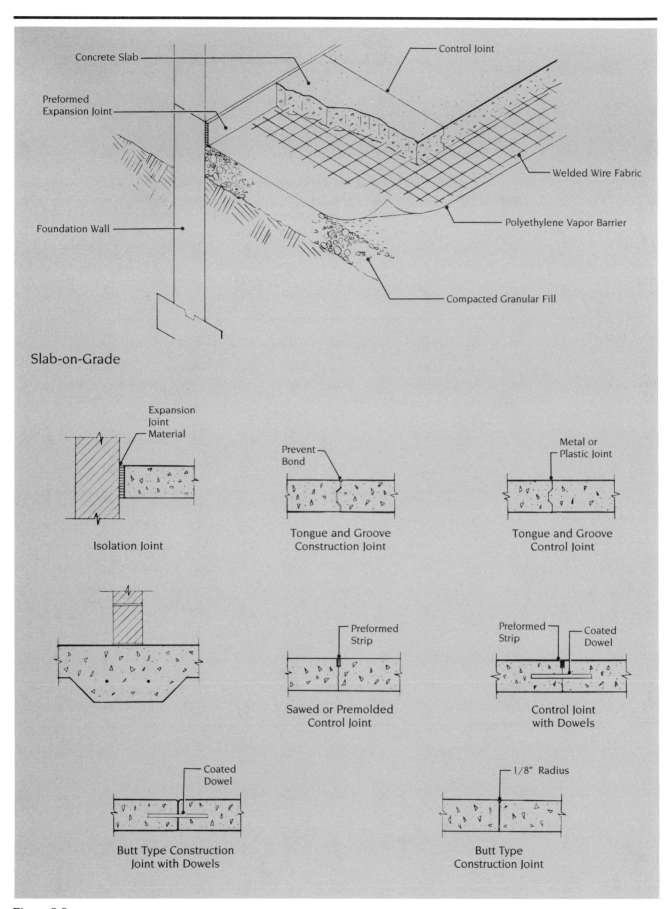

Concrete Slab

Control Joint

Preformed Expansion Joint

Welded Wire Fabric

Foundation Wall

Polyethylene Vapor Barrier

Compacted Granular Fill

Slab-on-Grade

Expansion Joint Material

Isolation Joint

Prevent Bond

Tongue and Groove Construction Joint

Metal or Plastic Joint

Tongue and Groove Control Joint

Preformed Strip

Sawed or Premolded Control Joint

Preformed Strip   Coated Dowel

Control Joint with Dowels

Coated Dowel

Butt Type Construction Joint with Dowels

1/8" Radius

Butt Type Construction Joint

Figure 3.3

17

SUPERSTRUCTURE	3.5-900	Mezzanine In Exist. Bldg.

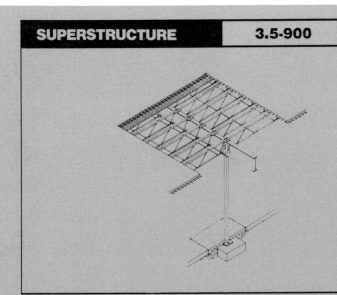

Mezzanine addition to existing building includes: Column footings; steel columns; structural steel; open web steel joists; uncoated 28 ga. steel slab forms; 2-1/2" concrete slab reinforced with welded wire fabric; steel trowel finish.

Design assumptions:
Structural steel is A36, high strength bolted. Slab form is 28 gauge, galvanized.

WWF 6 x 6 #10/#10
Conc. slab f'c = 3 ksi

System Components	QUANTITY	UNIT	COST EACH MAT.	COST EACH INST.	COST EACH TOTAL
SYSTEM 03.5-900					
MEZZANINE ADDITION TO EXISTING BUILDING; 100 PSF SUPERIMPOSED LOAD,					
SLAB FORM DECK, MTL. JOISTS, 2.5" CONC. SLAB, 3,000 PSI, 3,000 S.F.					
Concrete footing, 3' sq.	4.000	C.Y.	1,255.13	4,677.31	5,932.44
Column, 4" x 4" x ¼" x12'	12.000	Ea.	1,584	576	2,160
Structural steel w 21x50	80.000	L.F.	2,336.40	423.60	2,760
W 16x31	80.000	L.F.	1,578	384	1,962
Open web joists, H series	6.800	Ton	7,854	3,366	11,220
Slab form, steel 28 gauge, galvanized	30.000	C.S.F.	1,170	690	1,860
Concrete 3,000 psi, 2 ½" slab, incl. premium delv. chg.	23.500	C.Y.	1,974		1,974
Welded wire fabric 6x6 #10/#10 (w1.4/w1.4)	30.000	C.S.F.	257.40	492.60	750
Place concrete	23.500	C.Y.		740.25	740.25
Monolithic steel trowel finish	30.000	C.S.F.		1,470	1,470
Curing with sprayed membrane curing compound	30.000	C.S.F.	56.10	128.40	184.50
TOTAL			18,065.03	12,948.16	31,013.19
COST per S.F.			6.02	4.32	10.34

Figure 3.4

Chapter 4
MASONRY

This chapter contains descriptions of masonry work often specified for interior projects, including sizes and types available, various applications, and installation methods. Applications for masonry on interior projects are usually limited to finish materials on floors and walls, however, the masonry portion of an interior project can be very expensive. Materials, such as marble and granite, used in lobbies and foyers of commercial buildings, restaurant, retail wall or floor detailing, and other high visibility areas, can be costly. Installation of these items is a highly specialized trade, and can also be expensive. The estimator should determine the scope of masonry required for the project, and should obtain bids from specialty contractors.

Brick   Brick is a popular accent used for interior walls. It is available in most areas in varied sizes and colors and may be laid up in various bonds, jointing, and mortar colors. Common size brick measures nominal 4″ wide x 8″ long with heights of 2-2/3″ (standard), 3-1/5″ (engineer), 4″ (economy), 5-1/3″ (double), and 8″ (panel). Bricks 4″ wide x 12″ long are available in heights of 2″ (Roman), 2-2/3″ (Norman), 3-1/5″ (Norwegian), 4″ (utility), 5-1/3″ (triple), and 12″ (square or panel). The height of brick courses may also vary with the thickness of the mortar joints. Commonly used mortar joints vary from 3/8″ to 1/2″. A chart of brick sizes and corresponding mortar quantities required for various brick bond patterns is shown in Figure 4.1.

Bricks are installed in many different patterns within the wall to enhance its appearance. The six basic patterns of laying brick are stretcher, header, soldier, sailor, rowlock, and shiner. The various combinations of stretcher and header patterns within brick walls determine its bond. *Running Bond* consists of stretchers with no headers and staggered courses. Stretchers lined up vertically makes up *Stack Bond. Common Bond* is made up of five courses of stretchers and one course of headers. *Flemish Bond* is composed of stretchers and headers laid alternately in the same course. *English Bond* consists of alternating courses of stretchers and headers. Originally, full headers were used to attach the brick to a masonry back-up wall or another wythe of bricks. This is now usually accomplished by the use of metal, corrugated brick wall ties, and the headers are usually clipped or half bricks. Figures 4.2 and 4.3 illustrate various brick patterns and bonds.

The proportional content of the mortar for all types of masonry is very important, especially in bearing partitions. The chart in Figure 4.4 illustrates various types of mortar with appropriate quantities and characteristics.

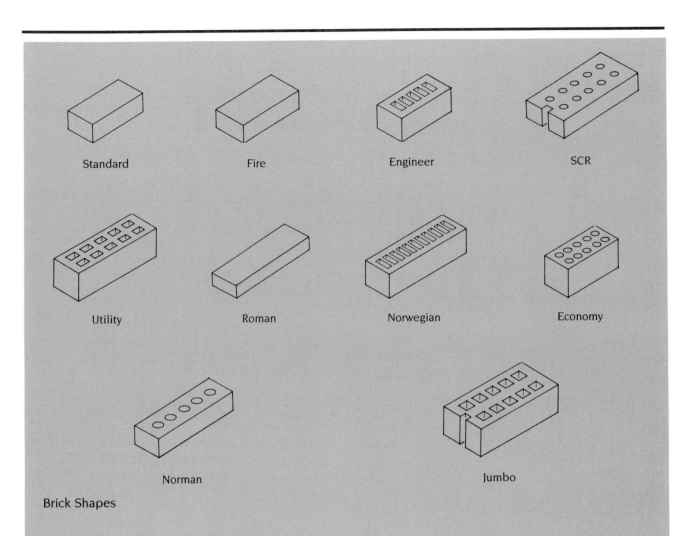

Brick Shapes

Brick & Mortar Quantities

| | Running Bond | | | | | For Other Bonds Standard Size Add to S F Quantities in Table to Left | | |
| | Number of Brick per SF of Wall - Single Wythe with 1/2" Joints | | | CF of Mortar per M Bricks, Waste Included | | | | |
Type Brick	Nominal Size (Incl. Mortar) L H W	Modular Coursing	Number of Brick per SF	1 Wythe	2 Wythe	Bond Type	Description	Factor
Standard	8 x 2⅔ x 4	3C x 8"	6.75	12.9	16.5	Common	full header every fifth course	+20%
Economy	8 x 4 x 4	1C = 4"	4.50	14.6	19.6		full header every sixth course	+16.7%
Engineer	8 x 3⅕ x 4	5C = 16"	5.63	13.6	17.6	English	full header every second course	+50%
Fire	9 x 2½ x 4½	2C = 5"	6.40	550# fireclay	—	Flemish	alternate headers every course	+33.3%
Jumbo	12 x 4 x 6 or 8	1C = 4"	3.00	34.0	41.4		every sixth course	+5.6%
Norman	12 x 2⅔ x 4	3C = 8"	4.50	17.8	22.8	Header = W x H exposed		+100%
Norwegian	12 x 3⅕ x 4	5C = 16"	3.75	18.5	24.4	Rowlock = H x W exposed		+100%
Roman	12 x 2 x 4	2C = 4"	6.00	17.0	20.7	Rowlock stretcher = L x W exposed		+33.3%
SCR	12 x 2⅔ x 6	3C = 8"	4.50	26.7	31.7	Soldier = H x L exposed		—
Utility	12 x 4 x 4	1C = 4"	3.00	19.4	26.8	Sailor = W x L exposed		–33.3%

Figure 4.1

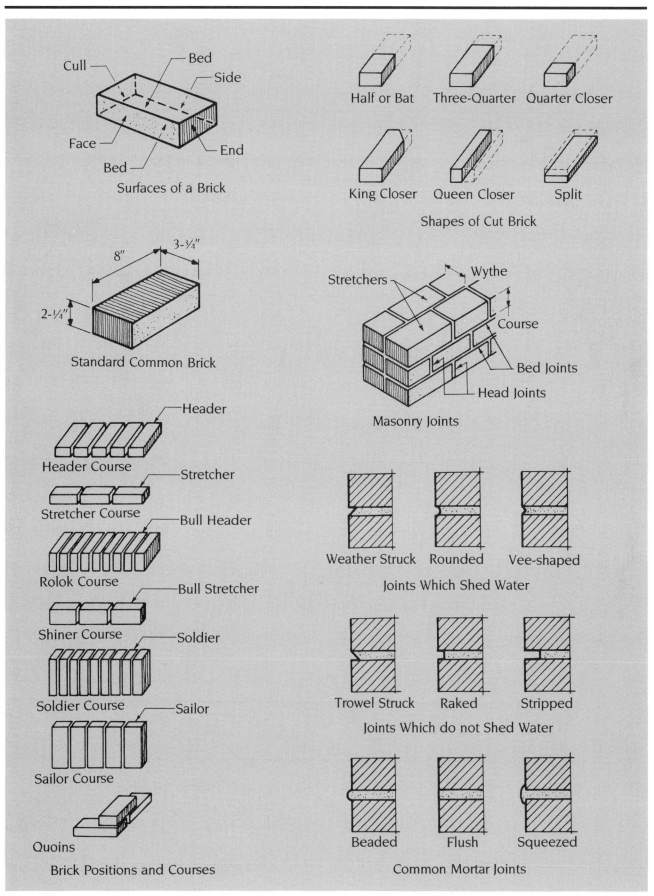

Surfaces of a Brick

Half or Bat   Three-Quarter   Quarter Closer

King Closer   Queen Closer   Split

Shapes of Cut Brick

Standard Common Brick

8"   3-¾"   2-¼"

Stretchers   Wythe   Course   Bed Joints   Head Joints

Masonry Joints

Header — Header Course

Stretcher — Stretcher Course

Bull Header — Rolok Course

Bull Stretcher — Shiner Course

Soldier — Soldier Course

Sailor — Sailor Course

Quoins

Brick Positions and Courses

Weather Struck   Rounded   Vee-shaped

Joints Which Shed Water

Trowel Struck   Raked   Stripped

Joints Which do not Shed Water

Beaded   Flush   Squeezed

Common Mortar Joints

Figure 4.2

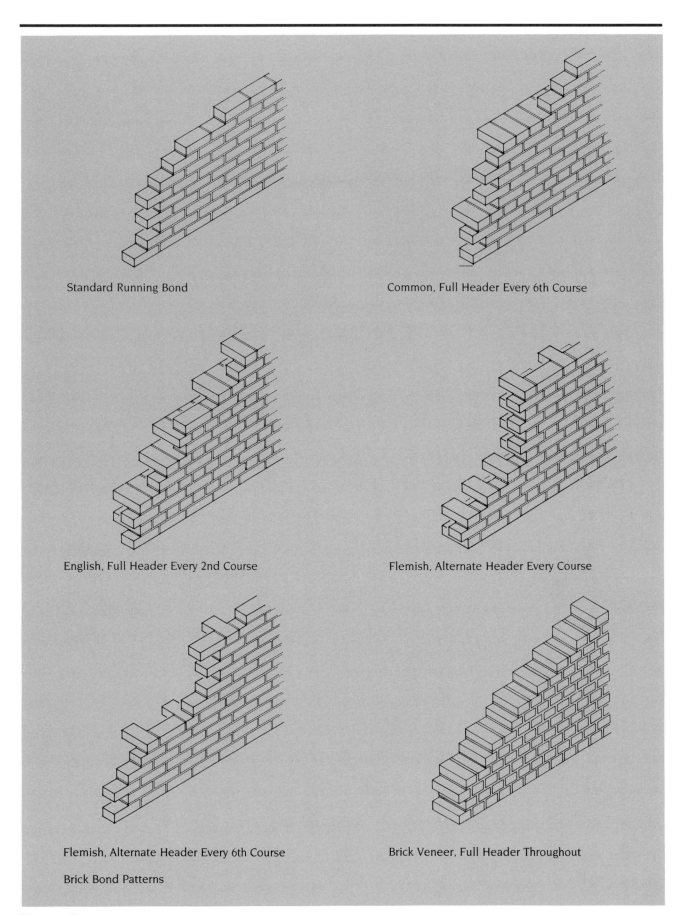

Standard Running Bond

Common, Full Header Every 6th Course

English, Full Header Every 2nd Course

Flemish, Alternate Header Every Course

Flemish, Alternate Header Every 6th Course

Brick Veneer, Full Header Throughout

Brick Bond Patterns

Figure 4.3

Brick Mortar Mixes*					
Type	Portland Cement	Hydrated Lime	Sand** (Maximum)	Strength	Use
M	1	1/4	3-3/4	High	General use where high strength is required, especially good compressive strength; work that is below grade and in contact with earth.
S	1	1/2	4-1/2	High	Okay general for use, especially good where high lateral strength is desired.
N	1	1	6	Medium	General use when masonry is exposed above grade; best to use when high compressive and lateral strengths are not required.
0	1	2	9	Low	Do not use when masonry is exposed to severe weathering; acceptable for non-loadbearing walls of solid units and interior non-loadbearing partitions of hollow units.

*The water used should be of the quality of drinking water. Use as much water as is needed to bring the mix to a suitably plastic and workable state.

**The sand should be damp and loose. A general rule for sand content is that it should not be less than 2-1/4 or more than 3 times the sum of the cement and lime volumes.

Figure 4.4

Concrete Block

For reasons of strength, versatility, and economy, concrete blocks are frequently used for constructing masonry walls and partitions. They may be used for interior bearing walls, infill panels, interior partitions, and fire walls. If the concrete wall is to serve as the finished wall, many architectural finishes are available, including ground (exposed aggregate), split, scored, and split-ribbed surfaces. Slump block, which features a distinctive bulging face, and blocks with geometric faces or embossed patterns, provide other options for enhancing the appearance of exposed block-wall surfaces. Applications of various types of concrete block are shown in Figures 4.5a and 4.5b. Typical block shapes are shown in Figure 4.6.

Concrete blocks are manufactured in two basic types, *solid* and *hollow*, and in various strength ratings. If the cross-sectional area, exclusive of voids, is 75% or greater than the gross area of the block, then it is classified as *solid block*. If the area is below the 75% figure, then the block is classified as *hollow block*. The strength of concrete block is determined by the compressive strength of the type of concrete used in its manufacture, or by the equivalent compressive strength, which is based on the gross area of the block, including voids.

There are several special aggregates that can be used to manufacture lightweight blocks. These blocks can be identified by the weight of the concrete mixture used in their manufacture. Regular weight block is made from 125 lb. per cubic foot (PCF) concrete and lightweight block from 105 to 85 PCF concrete. Costs for lightweight concrete as compared to regular weight range from 20% more for 105 PCF block to 50% more for 85 PCF block. Blocks of various strengths, grades, finishes, and weights should be taken off separately as costs may vary considerably.

Joint reinforcing and individual ties serve as important components of the various types of concrete block walls and must be included in the estimate. Two types of joint reinforcing are available: the *truss type* and the *ladder type* (shown in Figure 4.7). Because the truss type provides better load distribution, it is normally used in bearing walls. The ladder type is usually installed in light-duty walls that serve non-bearing functions. Both types of joint reinforcing may also be used to tie together the inner and outer wythes of composite or cavity-design walls. Corrugated strips, as well as Z-type, rectangular, and adjustable wall ties, may also be used for this purpose. Generally, one metal wall tie should be installed for each 4-1/2 square feet of veneer. Although both types of joint reinforcing may be used as ties, individual ties should not be used as joint reinforcing to control cracking.

Structural reinforcement is commonly required in concrete block walls, especially in those that are loadbearing. Deformed steel bars may be used as vertical reinforcement when grouted into the block voids, and as horizontal reinforcement when installed above openings and in bond beams. Horizontal and vertical bars may be grouted into the void normally used as the collar joint in a composite wall. Lintels should be installed to carry the weight of the wall above openings. Steel angles, built-up steel members, bond beams filled with steel bars and grout, and precast shapes may function as lintels.

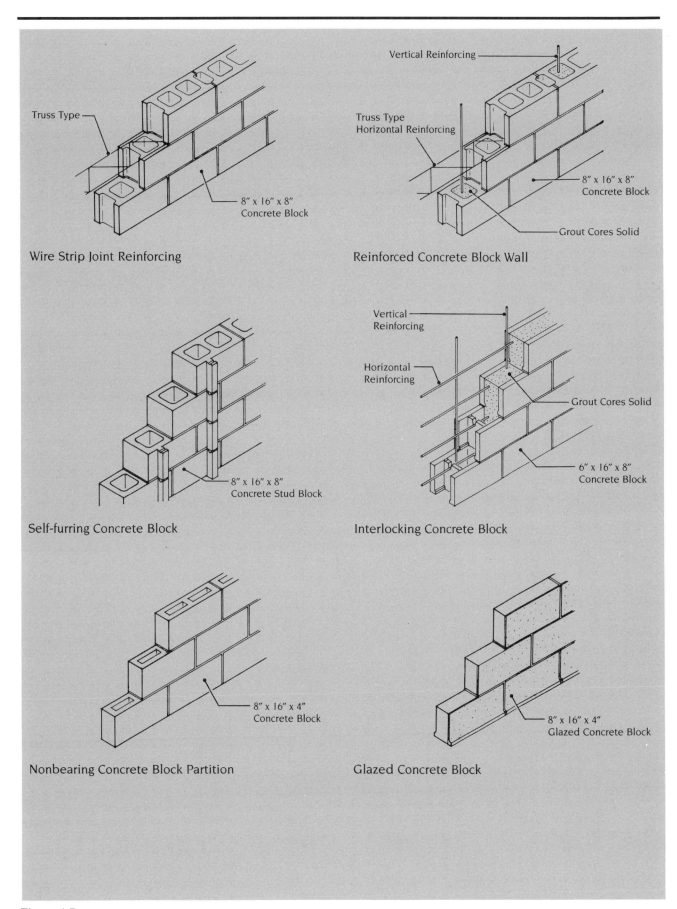

Wire Strip Joint Reinforcing

Reinforced Concrete Block Wall

Self-furring Concrete Block

Interlocking Concrete Block

Nonbearing Concrete Block Partition

Glazed Concrete Block

Figure 4.5a

25

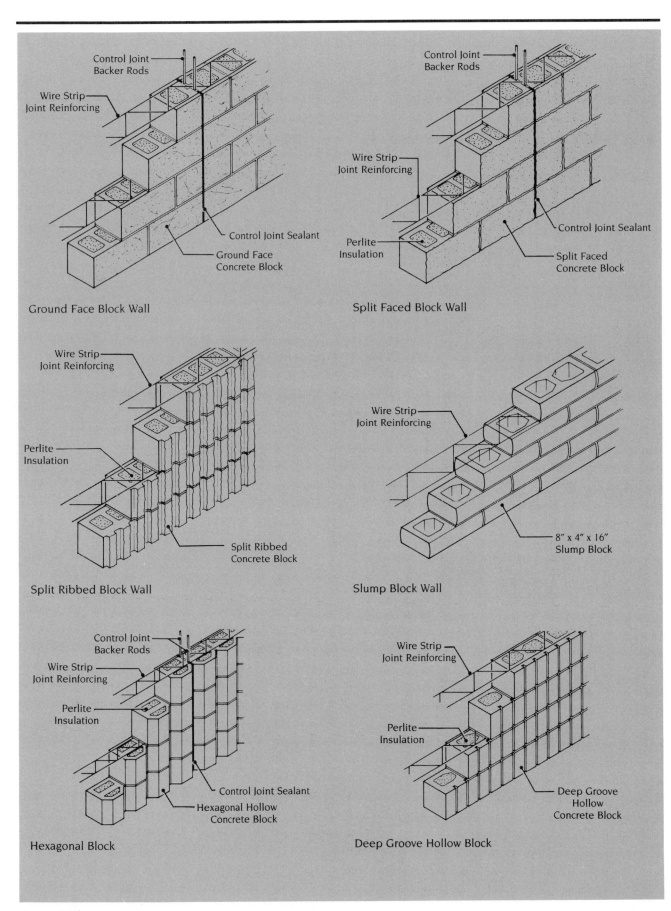

Ground Face Block Wall

Split Faced Block Wall

Split Ribbed Block Wall

Slump Block Wall

Hexagonal Block

Deep Groove Hollow Block

Figure 4.5b

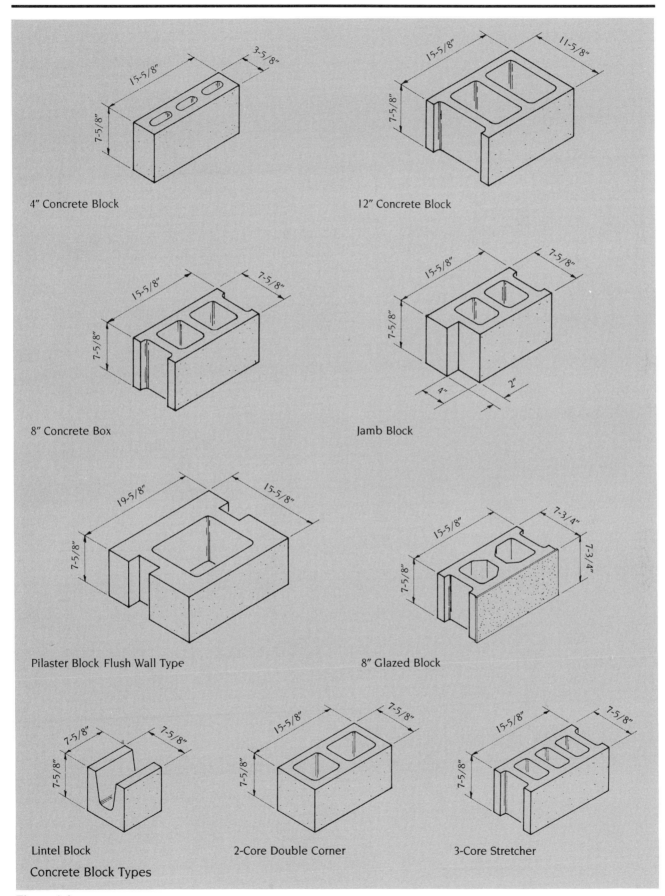

4" Concrete Block

12" Concrete Block

8" Concrete Box

Jamb Block

Pilaster Block Flush Wall Type

8" Glazed Block

Lintel Block

2-Core Double Corner

3-Core Stretcher

Concrete Block Types

Figure 4.6

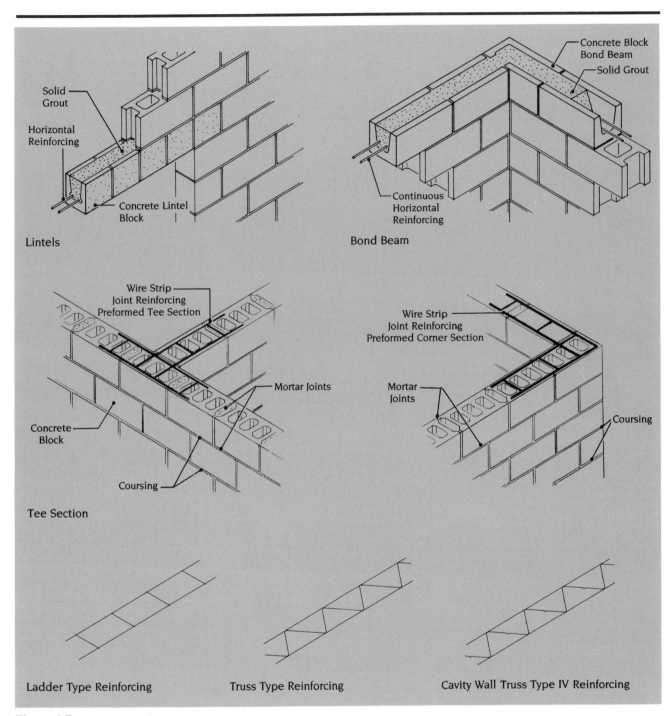

Lintels

Bond Beam

Tee Section

Ladder Type Reinforcing   Truss Type Reinforcing   Cavity Wall Truss Type IV Reinforcing

Figure 4.7

Concrete block partitions may be erected of nominal 4″, 6″, 8″, 10″, or 12″ thick concrete masonry units. The units may be regular weight or lightweight and solid or 75% solid in horizontal profile. Normal units are nominal 16″ long but some manufacturers produce a nominal 24″ unit. They may be left exposed, painted, or epoxy-coated, or furred and then covered with gypsum board or paneling. Partitions may also be plastered (directly on the block) or covered with self-furring lath and then plastered. Figure 4.7 shows various types of concrete block specialties: lintels, corners, bond beam, and joint reinforcing. Figure 4.8 shows a page from Means *Interior Cost Data*, 1987 illustrating how to estimate concrete block partitions as a system.

Structural Facing Tile

Structural glazed facing tile (SGFT) is kiln-fired structural clay with an integral impervious ceramic face. It may also be manufactured with an acoustical perforated face. Because glazed tile resists stains, marks, impact, abrasion, fading, and cracking, it is ideally suited for use in school corridors, locker rooms, rest rooms, kitchens, and other places where cleanliness and indestructibility are primary considerations.

Structural glazed facing tile is commonly available in a large selection of colors and color combinations in the 6T series, with 5-1/3′ x 12″ nominal face and in the 8W series, with 8″ x 16″ nominal face, both in 2″, 4″, 6″, and 8″ widths. Some manufacturers produce a 4W series with 8″ x 8″ nominal face in 2″ and 4″ widths.

Some available tile shapes include stretchers, bullnose jamb or corner, square jamb or corner, covered internal corner, recessed cove base, non-recessed cove base, bullnose sill, square sill, and universal miter. Walls with openings and returns usually require partition layout drawings to establish quantities of special shapes. The different shapes of structural glazed facing tile are shown in Figure 4.9.

Building Stone

For reasons of durability and unique appearance, stone provides a wide range of decorative applications as a building material. It can be installed in small units, referred to as "building stone", which can be assembled in many different formats, with or without mortar, to create walls and veneers of all sorts. Stone can also be employed as a material in larger units, such as stone panels, which are installed with elaborate anchor and framing systems and used as decorative wall facings in high-rise office and other commercial buildings. The cost of stone building materials varies considerably with location and depends on the available supply and the distance that it must be transported. Other significant cost factors include the extent of quarrying and subsequent processing required to extract and produce the finished product.

Building stone may be used in random or pre-cut sizes and shapes. Small, irregular building stone that has been quarried in random sizes is called "rubble" or "fieldstone". This material, which is sold by the ton, is commonly installed with mortar. Fieldstone may be split by hand on site to provide a flat exterior surface for a patterned wall or fireplace.

The Concrete Block Partition Systems are defined by weight and type of block, thickness, type of finish and number of sides finished. System components include joint reinforcing on alternate courses and vertical control joints.

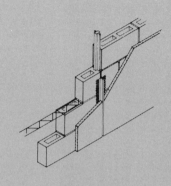

System Components		QUANTITY	UNIT	COST PER S.F.		
				MAT.	INST.	TOTAL
SYSTEM 06.1-210-1020						
CONC. BLOCK PARTITION, 8″ X 16″, 4″ TK., 2 CT. GYP. PLASTER 2 SIDES						
Conc. block partition, 4″ thick		1.000	S.F.	.86	2.51	3.37
Control joint		.050	L.F.	.08	.06	.14
Horizontal joint reinforcing		.800	L.F.	.08	.02	.10
Gypsum plaster, 2 coat, on masonry		2.000	S.F.	.49	2.50	2.99
	TOTAL			1.51	5.09	6.60

6.1-210		Concrete Block Partitions - Regular Weight						
		THICKNESS				COST PER S.F.		
	TYPE	(IN.)	TYPE FINISH	SIDES FINISHED		MAT.	INST.	TOTAL
1000	Hollow	4	none	0		1.02	2.59	3.61
1010			gyp. plaster 2 coat	1		1.27	3.84	5.11
1020				2		1.51	5.10	6.61
1050			fiber plaster - 2 coat	1		1.28	4.33	5.61
1100			lime plaster - 2 coat	1		1.15	3.84	4.99
1150			lime portland - 2 coat	1		1.14	3.84	4.98
1200			portland - 3 coat	1		1.08	4.03	5.11
1400			⅝″ drywall	1		1.54	3.54	5.08
1410				2		1.86	3.82	5.68
1500		6	none	0		1.09	2.78	3.87
1510			gyp. plaster 2 coat	1		1.34	4.03	5.37
1520				2		1.58	5.30	6.88
1550			fiber plaster - 2 coat	1		1.35	4.52	5.87
1600			lime plaster - 2 coat	1		1.22	4.03	5.25
1650			lime portland - 2 coat	1		1.21	4.03	5.24
1700			portland - 3 coat	1		1.15	4.22	5.37
1900			⅝″ drywall	1		1.61	3.73	5.34
1910				2		2.13	4.68	6.81
2000		8	none	0		1.30	2.97	4.27
2010			gyp. plaster 2 coat	1		1.55	4.22	5.77
2020			gyp. plaster 2 coat	2		1.79	5.45	7.24
2050			fiber plaster - 2 coat	1		1.56	4.71	6.27

274

Figure 4.8

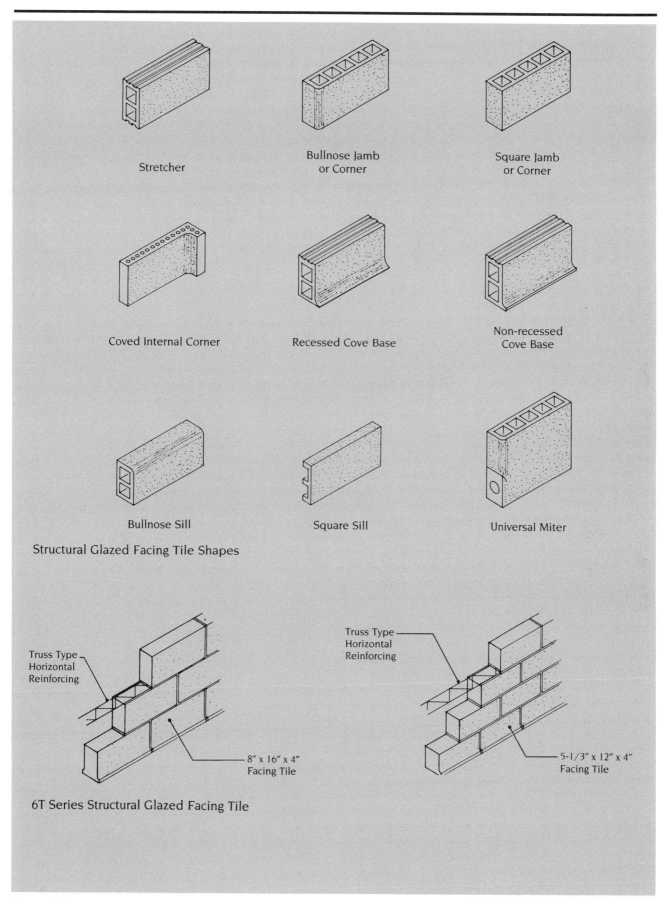

Stretcher

Bullnose Jamb
or Corner

Square Jamb
or Corner

Coved Internal Corner

Recessed Cove Base

Non-recessed
Cove Base

Bullnose Sill

Square Sill

Universal Miter

Structural Glazed Facing Tile Shapes

Truss Type
Horizontal
Reinforcing

8" x 16" x 4"
Facing Tile

Truss Type
Horizontal
Reinforcing

5-1/3" x 12" x 4"
Facing Tile

6T Series Structural Glazed Facing Tile

Figure 4.9

Small stone units may also be quarry split and processed to meet aesthetic needs. For example, decorative building stone can be purchased by the ton in 4″ thick slabs, available in lengths ranging from 6″ to 14″ and in heights ranging from 2″ to 16″. These pieces are commonly installed with mortar to create veneer walls of varying patterns, such as ledge stone, spider web, uncoursed rectangular, and squared. Ashlar stone, also priced by the ton, is building stone that has been sawn on the edges so as to produce a rectangular face. This shape makes ashlar stone another possible veneer material because the pieces can be arranged in either a regular or random-coursed pattern within the face of a wall. Typical patterns are shown in Figure 4.10. Stone veneer can be tied to the backup wall with galvanized ties or 8″ stone headers in a method similar to that used in brick veneer walls. The coverage of stone veneer ranges from 35 to 50 square feet per ton for 4″ wide veneer, with correspondingly reduced coverages per ton for veneers of 6″ and 8″ in width. Figure 4.11 is a page from Means *Interior Cost Data*, 1987, showing a typical stone veneer system.

Large stone facing materials can be installed as decorative features in many types of commercial buildings. These panels, which are usually priced by the square foot, are available in widths of up to 5′ and in thicknesses of approximately 1″ to 5″. Panel faces may be clear or patterned with split, sawn, or sand-rubbed surface finishes. The edges of the panel are saw cut and the back is planed.

Stone Floors

Stone floors may be constructed from any type of stone that meets the durability standards and aesthetic requirements of their locations. Some of the commonly used stone flooring materials include slate, flagstone, granite, and marble. The floors may adopt patterned or random designs and use stones that are randomly cut, uniform in size and shape, or designed in patterned sets. The stones may feature neat sawn edges or the irregular shapes of field-cut edges. The various possible exposed face finishes for the stones include: natural cleft, sawn, sawn and polished, and any specialized finish available to the type of stone which is functionally and aesthetically appropriate.

Stone floors may be placed on mortar beds or applied on mastic adhesive. When the floor is installed on a mortar bed, the stone may vary slightly in thickness, with the total floor usually measuring 1-1/2″ to 2″ thick from subfloor to finish. When the stone floor is thin set, or laid in mastic, gauged stones with a constant thickness must be used. After the stone flooring material has been placed, pre-mixed grouting material in various colors or mortar is used to fill the joints between the stones. Some stone flooring materials with consistent sizes and regular edges may not require grouted joints. Many different sealants for the various stone types are available for the final step in the installation process. Some of the stone patterns available are shown in Figure 4.12.

Patterns of Stonework

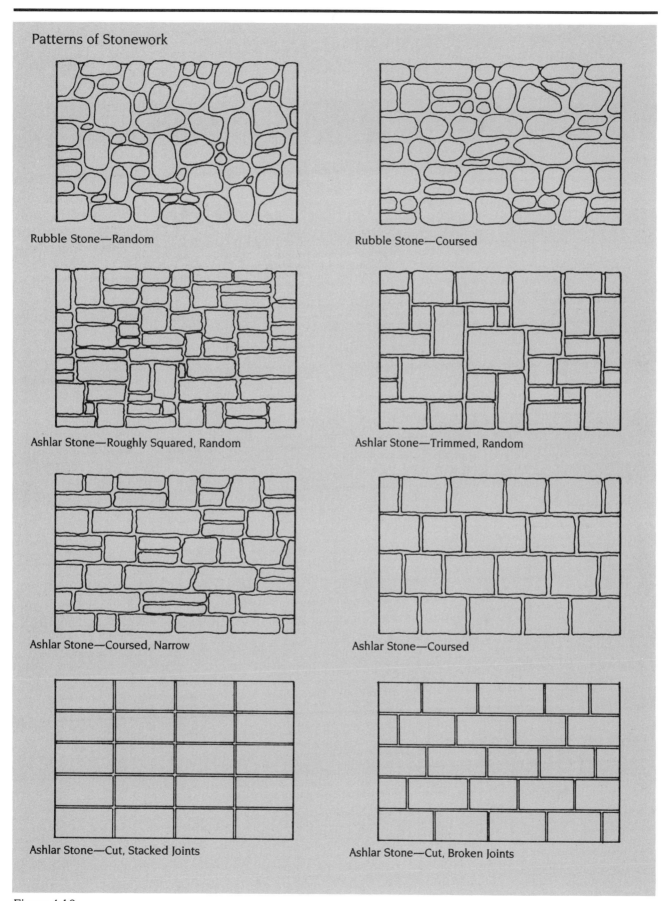

Rubble Stone—Random

Rubble Stone—Coursed

Ashlar Stone—Roughly Squared, Random

Ashlar Stone—Trimmed, Random

Ashlar Stone—Coursed, Narrow

Ashlar Stone—Coursed

Ashlar Stone—Cut, Stacked Joints

Ashlar Stone—Cut, Broken Joints

Figure 4.10

BUILDING SUBDIVISION	6.1-242	Stone Veneer

Systems are listed by costs per S.F. for stone veneer walls on various backup using different stone. Typical components for a system are shown in the component block below.

System Components	QUANTITY	UNIT	COST PER S.F.		
			MAT.	INST.	TOTAL
SYSTEM 06.1-242-2000					
ASHLAR STONE VENEER 4″, 2″X4″ STUD 16″ O.C.BACK UP, 8′ HIGH					
Ashlar veneer, 4″ thick, sawn face, split joints, low priced stone	1.000	S.F.	4.96	8.99	13.95
Partitions, 2″ x 4″ studs 8′ high 12″ O.C.	.920	B.F.	.33	.63	.96
Wall ties for stone veneer galv. corrg ⅞″ x 7″, 24 gage	.700	Ea.	.02	.16	.18
Sheathing plywood on wall, CDX ½″	1.000	S.F.	.43	.45	.88
TOTAL			5.74	10.23	15.97

6.1-242	Stone Veneer	COST PER S.F.		
		MAT.	INST.	TOTAL
2000	Ashlar stone veneer,4″ thick, $110/ton,wood stud backup,8′ high,16″ O.C.	5.75	10.25	16
2100	Metal stud backup, 8′ high, 16″ O.C.	6.50	10.20	16.70
2150	24″ O.C.	6.35	10.15	16.50
2200	Conc. block backup, 4″ thick	6.30	12.05	18.35
2300	6″ thick	6.50	12.15	18.65
2350	8″ thick	6.70	12.30	19
2400	10″ thick	7.30	12.35	19.65
2500	12″ thick	7	13	20
3100	$250/ton, wood stud backup, 10′ high, 16″ O.C.	11.55	10.25	21.80
3200	Metal stud backup, 10′ high, 16″ O.C.	12.35	10.25	22.60
3250	24″ O.C.	12.15	10.15	22.30
3300	Conc. block backup, 10′ high, 4″ thick	12.15	12.10	24.25
3350	6″ thick	12.30	12.10	24.40
3400	8″ thick	12.55	12.35	24.90
3450	10″ thick	13.10	12.40	25.50
3500	12″ thick	12.85	13.05	25.90
4000	Indiana limestone 2″ thick,sawn finish,wood stud backup,10′ high,16″ O.C.	10.90	12.10	23
4100	Metal stud backup, 10′ high, 16″ O.C.	11.70	12.10	23.80
4150	24″ O.C.	11.50	12.05	23.55
4250	6″ thick	11.65	14.05	25.70
4300	8″ thick	11.85	14.20	26.05
4350	10″ thick	12.45	14.25	26.70
4400	12″ thick	12.15	14.90	27.05
4450	2″ thick, smooth finish, wood stud backup, 8′ high, 16″ O.C.	12.75	11.25	24
4550	Metal stud backup, 8′ high, 16″ O.C.	13.55	11.15	24.70
4600	24″ O.C.	13.35	11.05	24.40
4650	Conc. block backup, 4″ thick	13.35	13	26.35
4700	6″ thick	13.50	13.10	26.60
4750	8″ thick	13.75	13.25	27
4800	10″ thick	14.35	13.40	27.75

280

Figure 4.11

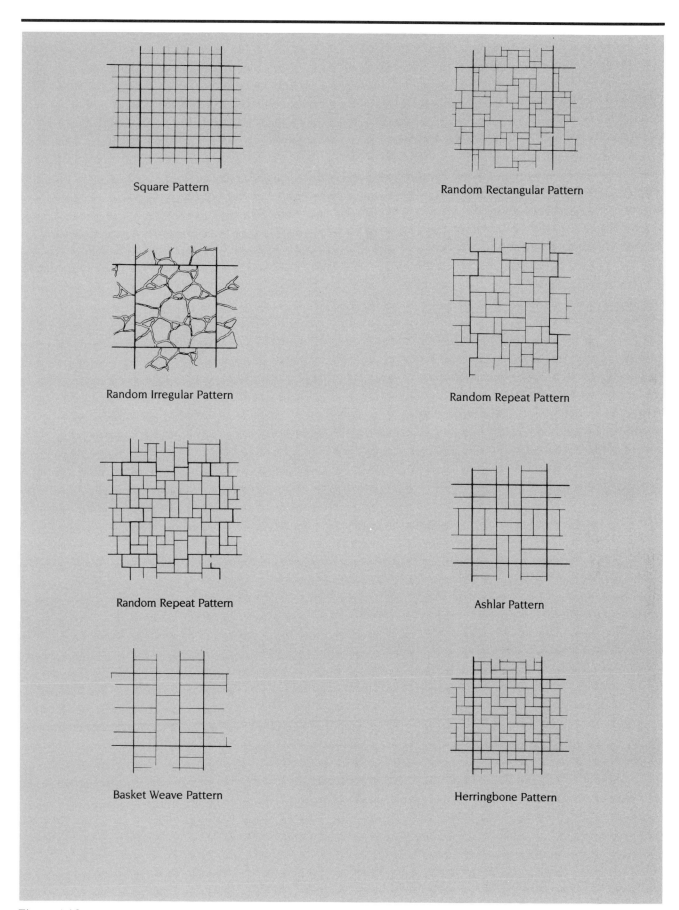

Square Pattern

Random Rectangular Pattern

Random Irregular Pattern

Random Repeat Pattern

Random Repeat Pattern

Ashlar Pattern

Basket Weave Pattern

Herringbone Pattern

Figure 4.12

Glass Block

Glass block, which was a popular building material in the 1930's, has recently undergone a revival of usage in the construction industry, particularly for the construction of interior partitions. It has the dual advantage of admitting light while providing privacy.

Glass block can be placed on a raised base, plate, or sill, provided that the surfaces to be mortared are primed with asphalt emulsion. Wall recesses and channel track that receive the glass block should be lined with expansion strips prior to oakum filler and caulking. Horizontal joint reinforcing is specified for flexural as well as shrinkage control and is laid in the joints along with the mortar. End blocks are anchored to the adjacent construction with metal anchors, if no other provisions for attachment exist. If intermediate support is required, vertical I-shaped stiffeners can either be installed in the plane of the wall or adjacent to it, but the stiffeners should be tied to the wall with wire anchors. The top of the wall is supported between angles or in a channel track similar to the jambs. Typical construction details for glass block are shown in Figure 4.13.

Glass block is manufactured in sizes from 6″ by 6″ to 12″ by 12″, and thicknesses from 3″ to 4″. The block may be hollow or fused brick; the latter allows vision through the block. Inserts can be manufactured into the block to reduce solar transmission. Quantities that can be used for estimating glass block are shown in Figure 4.14.

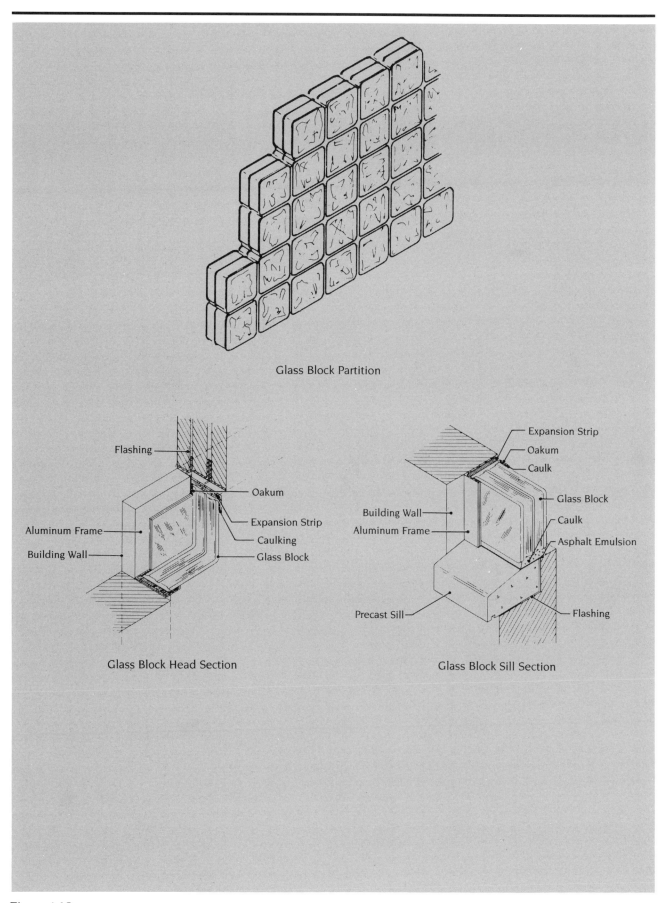

Glass Block Partition

Glass Block Head Section

Flashing
Oakum
Expansion Strip
Caulking
Aluminum Frame
Building Wall
Glass Block

Glass Block Sill Section

Expansion Strip
Oakum
Caulk
Glass Block
Building Wall
Caulk
Aluminum Frame
Asphalt Emulsion
Precast Sill
Flashing

Figure 4.13

37

Glass Block

Cost of blocks each, all blocks 4" thick, zone 1 contractor prices.

Nominal Size (Including Mortar)	Truckload or Carload				Less Than Truckload or Carload			
	Regular	Thinline	Essex	Solar Reflective	Regular	Thinline	Essex	Solar Reflective
6" x 6"	$ 3.20	$2.90	—	—	$ 3.80	$3.85	—	—
8" x 8"	4.70	3.40	$ 4.95	$ 8.40	5.80	5.40	$ 5.40	$ 9.00
Solid 8" x 8"	18.85	—	—	—	—	—	—	—
4" x 12"	—	—	—	—	—	—	—	—
4" x 8"	3.10	2.45	—	—	3.85	3.40	—	—
12" x 12"	11.90	—	—	—	12.90	—	—	—

Size	Per 100 S.F.		Per 1000 Block				
	No. of Block	Mortar 1/4" Joint	Asphalt Emulsion	Caulk	Expansion Joint	Panel Anchors	Wall Mesh
6" x 6"	410 ea.	5.0 C.F.	.17 gal.	1.5 gal.	80 L.F.	20 ea.	500 L.F.
8" x 8"	230	3.6	.33	2.8	140	36	670
12" x 12"	102	2.3	.67	6.0	312	80	1000
Approximate quantity per 100 S.F.			.07 gal.	.6 gal.	32 L.F.	9 ea.	51, 68, 102 L.F.

Figure 4.14

Chapter 5
METALS

The amount of structural steel used in an interior estimate is usually limited. Beams or plates may be employed to reinforce an existing floor system. Beams, channels, or angles may be used to frame a new opening or stairwell. The addition of a mezzanine floor might include more structural steel, requiring columns, beams, open web joists and steel deck, however, the most extensive use of metals in interior work is for miscellaneous support and ornamental purposes.

Structural Steel

Structural steel, when installed in an existing building, can be labor intensive. The steel may have to be moved into the building with dollies or hand carts, hoisted into position with chain falls or come-a-longs, raised with a fork lift, or jacked from a scaffold. New connections are either welded (if fire codes permit) or mag drilled and bolted. The installation process must be visualized and planned in order to properly estimate the costs. A "complexity factor" may be included for each piece of steel. Typical, available steel shapes are shown in Figure 5.1, along with drawing designations and typical connection details. Figures 5.2a through 5.2c show box and surface areas for available steel shapes. These quantities are useful when estimating fireproofing costs — whether sprayed or drywall type. A composite mezzanine system including the footing, steel, open web joists, and concrete slab is shown in Chapter 3, "Concrete", Figure 3.4. Costs are shown per S.F. of the total system.

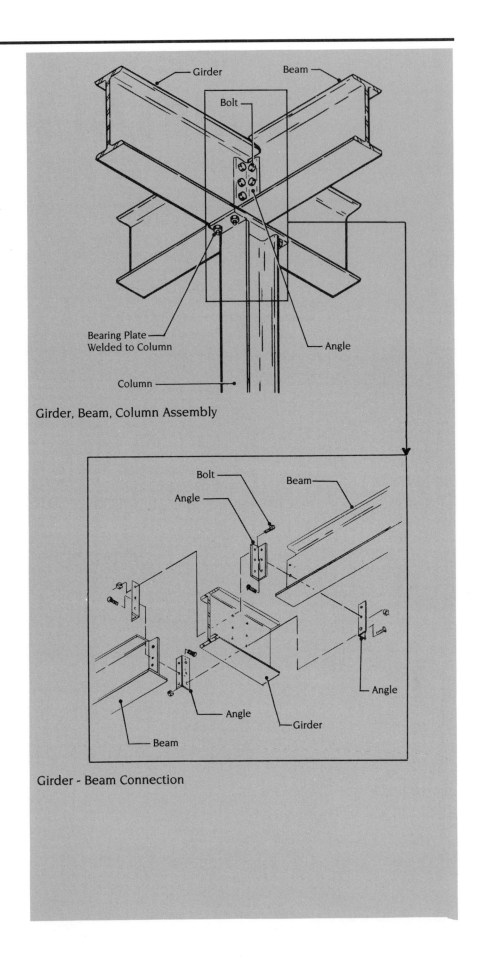

Girder, Beam, Column Assembly

Girder - Beam Connection

Figure 5.1

Surface Areas and Box Areas
W Shapes
Square feet per foot of length

Designation	Case A	Case B	Case C	Case D
W 36x300	9.99	11.40	7.51	8.90
x280	9.95	11.30	7.47	8.85
x260	9.90	11.30	7.42	8.80
x245	9.87	11.20	7.39	8.77
x230	9.84	11.20	7.36	8.73
x210	8.91	9.93	7.13	8.15
x194	8.88	9.89	7.09	8.10
x182	8.85	9.85	7.06	8.07
x170	8.82	9.82	7.03	8.03
x160	8.79	9.79	7.00	8.00
x150	8.76	9.76	6.97	7.97
x135	8.71	9.70	6.92	7.92
W 33x241	9.42	10.70	7.02	8.34
x221	9.38	10.70	6.97	8.29
x201	9.33	10.60	6.93	8.24
x152	8.27	9.23	6.55	7.51
x141	8.23	9.19	6.51	7.47
x130	8.20	9.15	6.47	7.43
x118	8.15	9.11	6.43	7.39
W 30x211	8.71	9.97	6.42	7.67
x191	8.66	9.92	6.37	7.62
x173	8.62	9.87	6.32	7.57
x132	7.49	8.37	5.93	6.81
x124	7.47	8.34	5.90	6.78
x116	7.44	8.31	5.88	6.75
x108	7.41	8.28	5.84	6.72
x 99	7.37	8.25	5.81	6.68
W 27x178	7.95	9.12	5.81	6.98
x161	7.91	9.08	5.77	6.94
x146	7.87	9.03	5.73	6.89
x114	6.88	7.72	5.39	6.23
x102	6.85	7.68	5.35	6.18
x 94	6.82	7.65	5.32	6.15
x 84	6.78	7.61	5.28	6.11
W 24x162	7.22	8.30	5.25	6.33
x146	7.17	8.24	5.20	6.27
x131	7.12	8.19	5.15	6.22
x117	7.08	8.15	5.11	6.18
x104	7.04	8.11	5.07	6.14
x 94	6.16	6.92	4.81	5.56
x 84	6.12	6.87	4.77	5.52
x 76	6.09	6.84	4.74	5.49
x 68	6.06	6.80	4.70	5.45
x 62	5.57	6.16	4.54	5.13
x 55	5.54	6.13	4.51	5.10

Surface Areas and Box Areas
W Shapes
Square feet per foot of length

Designation	Case A	Case B	Case C	Case D
W 21x147	6.61	7.66	4.72	5.76
x132	6.57	7.61	4.68	5.71
x122	6.54	7.57	4.65	5.68
x111	6.51	7.54	4.61	5.64
x101	6.48	7.50	4.58	5.61
x 93	5.54	6.24	4.31	5.01
x 83	5.50	6.20	4.27	4.96
x 73	5.47	6.16	4.23	4.92
x 68	5.45	6.14	4.21	4.90
x 62	5.42	6.11	4.19	4.87
x 57	5.01	5.56	4.06	4.60
x 50	4.97	5.51	4.02	4.56
x 44	4.94	5.48	3.99	4.53
W 18x119	5.81	6.75	4.10	5.04
x106	5.77	6.70	4.06	4.99
x 97	5.74	6.67	4.03	4.96
x 86	5.70	6.62	3.99	4.91
x 76	5.67	6.59	3.95	4.87
x 71	4.85	5.48	3.71	4.35
x 65	4.82	5.46	3.69	4.32
x 60	4.80	5.43	3.67	4.30
x 55	4.78	5.41	3.65	4.27
x 50	4.76	5.38	3.62	4.25
x 46	4.41	4.91	3.52	4.02
x 40	4.38	4.88	3.48	3.99
x 35	4.34	4.84	3.45	3.95
W 16x100	5.28	6.15	3.70	4.57
x 89	5.24	6.10	3.66	4.52
x 77	5.19	6.05	3.61	4.47
x 67	5.16	6.01	3.57	4.43
x 57	4.39	4.98	3.33	3.93
x 50	4.36	4.95	3.30	3.89
x 45	4.33	4.92	3.27	3.86
x 40	4.31	4.89	3.25	3.83
x 36	4.28	4.87	3.23	3.81
x 31	3.92	4.39	3.11	3.57
x 26	3.89	4.35	3.07	3.53

Figure 5.2a

Surface Areas and Box Areas W Shapes Square feet per foot of length						Surface Areas and Box Areas W Shapes Square feet per foot of length				
Designation	Case A	Case B	Case C	Case D		Designation	Case A	Case B	Case C	Case D
W 14x730	7.61	9.10	5.23	6.72		W 12x336	5.77	6.88	3.92	5.03
x665	7.46	8.93	5.08	6.55		x305	5.67	6.77	3.82	4.93
x605	7.32	8.77	4.94	6.39		x279	5.59	6.68	3.74	4.83
x550	7.19	8.62	4.81	6.24		x252	5.50	6.58	3.65	4.74
x500	7.07	8.49	4.68	6.10		x230	5.43	6.51	3.58	4.66
x455	6.96	8.36	4.57	5.98		x210	5.37	6.43	3.52	4.58
x426	6.89	8.28	4.50	5.89		x190	5.30	6.36	3.45	4.51
x398	6.81	8.20	4.43	5.81		x170	5.23	6.28	3.39	4.43
x370	6.74	8.12	4.36	5.73		x152	5.17	6.21	3.33	4.37
x342	6.67	8.03	4.29	5.65		x136	5.12	6.15	3.27	4.30
x311	6.59	7.94	4.21	5.56		x120	5.06	6.09	3.21	4.24
x283	6.52	7.86	4.13	5.48		x106	5.02	6.03	3.17	4.19
x257	6.45	7.78	4.06	5.40		x 96	4.98	5.99	3.13	4.15
x233	6.38	7.71	4.00	5.32		x 87	4.95	5.96	3.10	4.11
x211	6.32	7.64	3.94	5.25		x 79	4.92	5.93	3.07	4.08
x193	6.27	7.58	3.89	5.20		x 72	4.89	5.90	3.05	4.05
x176	6.22	7.53	3.84	5.15		x 65	4.87	5.87	3.02	4.02
x159	6.18	7.47	3.79	5.09		x 58	4.39	5.22	2.87	3.70
x145	6.14	7.43	3.76	5.05		x 53	4.37	5.20	2.84	3.68
x132	5.93	7.16	3.67	4.90		x 50	3.90	4.58	2.71	3.38
x120	5.90	7.12	3.64	4.86		x 45	3.88	4.55	2.68	3.35
x109	5.86	7.08	3.60	4.82		x 40	3.86	4.52	2.66	3.32
x 99	5.83	7.05	3.57	4.79		x 35	3.63	4.18	2.63	3.18
x 90	5.81	7.02	3.55	4.76		x 30	3.60	4.14	2.60	3.14
x 82	4.75	5.59	3.23	4.07		x 26	3.58	4.12	2.58	3.12
x 74	4.72	5.56	3.20	4.04		x 22	2.97	3.31	2.39	2.72
x 68	4.69	5.53	3.18	4.01		x 19	2.95	3.28	2.36	2.69
x 61	4.67	5.50	3.15	3.98		x 16	2.92	3.25	2.33	2.66
x 53	4.19	4.86	2.99	3.66		x 14	2.90	3.23	2.32	2.65
x 48	4.16	4.83	2.97	3.64						
x 43	4.14	4.80	2.94	3.61						
x 38	3.93	4.50	2.91	3.48						
x 34	3.91	4.47	2.89	3.45						
x 30	3.89	4.45	2.87	3.43						
x 26	3.47	3.89	2.74	3.16						
x 22	3.44	3.86	2.71	3.12						

Figure 5.2b

Surface Areas and Box Areas W Shapes Square feet per foot of length				
Designation	Case A	Case B	Case C	Case D
W 10x112	4.30	5.17	2.76	3.63
x100	4.25	5.11	2.71	3.57
x 88	4.20	5.06	2.66	3.52
x 77	4.15	5.00	2.62	3.47
x 68	4.12	4.96	2.58	3.42
x 60	4.08	4.92	2.54	3.38
x 54	4.06	4.89	2.52	3.35
x 49	4.04	4.87	2.50	3.33
x 45	3.56	4.23	2.35	3.02
x 39	3.53	4.19	2.32	2.98
x 33	3.49	4.16	2.29	2.95
x 30	3.10	3.59	2.23	2.71
x 26	3.08	3.56	2.20	2.68
x 22	3.05	3.53	2.17	2.65
x 19	2.63	2.96	2.04	2.38
x 17	2.60	2.94	2.02	2.35
x 15	2.58	2.92	2.00	2.33
x 12	2.56	2.89	1.98	2.31
W 8x 67	3.42	4.11	2.19	2.88
x 58	3.37	4.06	2.14	2.83
x 48	3.32	4.00	2.09	2.77
x 40	3.28	3.95	2.05	2.72
x 35	3.25	3.92	2.02	2.69
x 31	3.23	3.89	2.00	2.67
x 28	2.87	3.42	1.89	2.43
x 24	2.85	3.39	1.86	2.40
x 21	2.61	3.05	1.82	2.26
x 18	2.59	3.03	1.79	2.23
x 15	2.27	2.61	1.69	2.02
x 13	2.25	2.58	1.67	2.00
x 10	2.23	2.56	1.64	1.97
W 6x 25	2.49	3.00	1.57	2.08
x 20	2.46	2.96	1.54	2.04
x 15	2.42	2.92	1.50	2.00
x 16	1.98	2.31	1.38	1.72
x 12	1.93	2.26	1.34	1.67
x 9	1.90	2.23	1.31	1.64
W 5x 19	2.04	2.45	1.28	1.70
x 16	2.01	2.43	1.25	1.67
W 4x 13	1.63	1.96	1.03	1.37

Figure 5.2c

Miscellaneous and Ornamental Metals

Miscellaneous and ornamental metals may be used extensively in interior projects for both decorative and functional purposes. The purchase, fabrication, and erection of these items is a specialized trade and a reliable sub-bid should be obtained for this portion of the interior work. Following is a list of some of the items normally furnished and delivered by a miscellaneous metals supplier.

- Elevator shaft beam separators
- Angle sills with welded-on anchors
- Angle corner-guards with welded-on anchors
- Pipe bollards with welded-on base-plate anchors
- Cast-iron drain grates with frames
- Individual aluminum or steel sleeves for pipe or tube rails
- Cast or extruded abrasive metal nosings for concrete steps
- Templated sleeves welded on a steel flat for continuous pipe or tube-guardrails at balconies or roof
- Transformer vault door frames
- Malleable iron wedge inserts for attached or hung lintels
- Angle frames with welded-on anchors
- Elevator machine-room double-leaf aluminum floor hatches with compensating hinges
- Elevator machine-room ceiling hoist monorail beams
- Roof scuttles
- Stainless steel sleeves for swimming-pool ladders and guardrails
- Slab inserts for toilet partitions, operating room lights, and x-ray room ceiling supports
- Loose lintels, 12″ longer than the net opening
- Steel stairs, complete as shown on drawings including abrasive extruded nosings, if any, but excluding concrete fill or terrazzo
- Elevator shaft sill angles mounted on inserts, built-in channel door-jambs
- All gratings, including any support angles bolted to masonry or concrete or connected to inserts built in by others
- All open-riser ships or engineer's ladders, with diamond plate or grating treads
- Interior ladders to roof scuttles and under hatches
- Exterior ladders with goosenecks from low to high roofs, with or without safety cages as per drawings
- Steel bench supports
- Tube or pipe guardrails in steel, non-ferrous or stainless steel at balconies, roofs, and elsewhere as per drawings, including at interior concrete stairs
- Hung or attached angle lintels for brick supports connected to inserts built in by others
- Catwalks, strutted or suspended, complete with grating walkways, guardrails, and ladder accesses
- Spiral metal angle bases at wooden floor accesses, auditorium stage, book stack accesses, mezzanines, etc.
- Sheet metal angle bases in wooden floor rooms, such as gyms, interior raquetball, squash, wrestling stages, and prosceniums
- Wall handrails at ramps, places other than stairs, hospital hallways, etc.
- Stainless steel swimming pool ladders and pipe guardrails at bleachers
- Toilet partition supports in "Unistrut"
- X-ray machine supports in "Unistrut" or angles

- Operating and autopsy room light "spider-leg" supports
- Monitor supports throughout a hospital
- Hospital linear accelerator supports
- Rolling and fire partitions supports
- Computer room floor supports
- Proscenium grillages
- Acoustic baffle cloud supports
- Motor supports for overhead doors
- Entrance door and other door supports
- Welding of inserts on decking sheets for uses underneath, for ceiling supports
- Projection booth counterweighted port doors for fire protection
- Exterior door saddles, with or without Rixsons
- Interior floor transition door slip saddles
- Exterior door combination slip saddles
- Floor, wall, and ceiling non-ferrous expansion joints
- Folding partition supports
- Banquet hall movable partition supports
- Steel or aluminum louvres *not* in contact with any ductwork
- Ornamental metals for glass railings
- Ornamental metals for composite acrylic/wood railings
- Ornamental metals for combination panels and railings
- Ornamental metals for balusters, posts, trillage, and scroll railings
- Non-ferrous expansion joint covers
- Non-ferrous door saddles

Some of the items listed above are illustrated in Figure 5.3. The list of ornamental items included in any interior estimate may be extensive, and the takeoff and pricing can be time consuming. A reliable subcontractor should be contacted to provide a proper quotation.

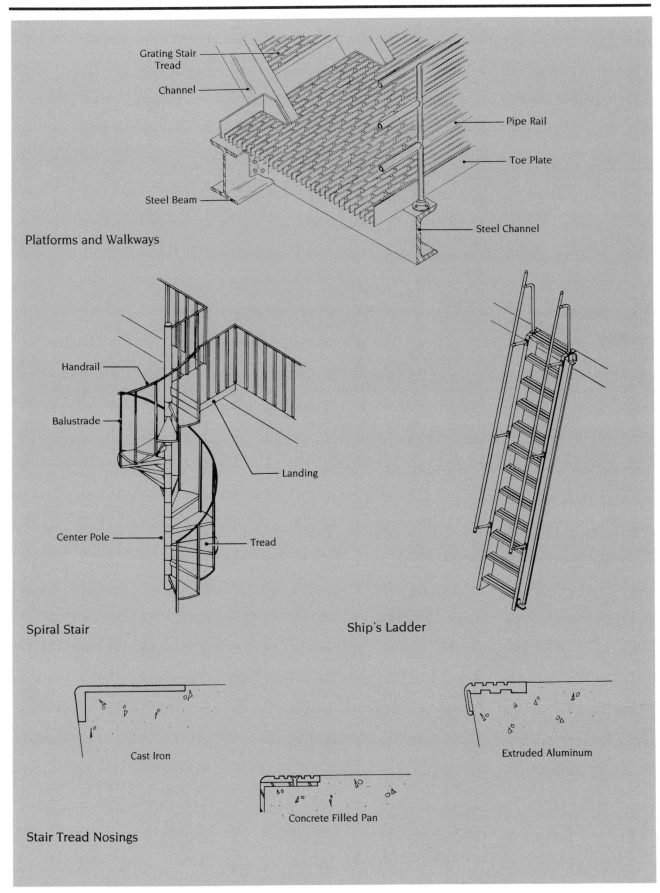

Grating Stair Tread

Channel

Pipe Rail

Toe Plate

Steel Beam

Steel Channel

Platforms and Walkways

Handrail

Balustrade

Landing

Center Pole

Tread

Spiral Stair

Ship's Ladder

Cast Iron

Extruded Aluminum

Concrete Filled Pan

Stair Tread Nosings

Figure 5.3

Chapter 6
WOOD AND PLASTICS

Wood, unlike most processed building materials, is an organic material that can be used in its natural state. Factors that influence its strength are density, natural defects (knots, grain, etc.), and moisture content. Wood can be easily shaped or cut to size on the site or prefabricated in the shop. Various framing systems may be used with fiberboard, particle board, plywood, or wood decking (plain or laminated) to form composite panels.

Material prices for lumber and wood products fluctuate more and with greater frequency than any other building material. For this reason, when the material list is complete it is important to obtain current, local prices for the lumber. Installation costs greatly depend on productivity. For Division 6, accurate cost records from past jobs can be most helpful, as lumber may be used extensively in interior projects.

Carpentry work can be broken down into the following categories: rough carpentry, finish carpentry and millwork, and laminated framing and decking. The rough carpentry materials can be sticks of lumber or sheets of plywood – job site fabricated and installed, or may consist of trusses, truss joints, and panelized roof systems – prefabricated, delivered, and erected by a specialty subcontractor.

Rough Carpentry

Rough carpentry may be required in interior projects where there is extensive renovation, or where location of openings, windows, or doors are significantly changed. Lumber is usually estimated in board feet and purchased in 1000 board foot quantities. A board foot is the equivalent of 1″ x 12″ x 12″ (nominal) or 3/4″ x 11-1/2″ x 12″ milled (actual). To determine board feet of a piece of framing, the nominal dimensions can be multiplied, and the result divided by 12. The final result represents the number of board feet per linear foot of that framing size.

Example: 2 x 10 joists
2 x 10 = 20 $\frac{20}{12} = 1.67 \frac{B.F.}{L.F.}$

The Quantity Sheet for lumber should indicate species, grade, and any type of wood preservative or fire retardant treatment specified or required by code. Floor joists, shown or specified by size and spacing, should be taken off by nominal length and the quantity required. Add for double joists under partitions, headers and cripple joists at openings, overhangs, laps at bearings, and blocking or bridging.

Studs required are noted on the drawings by spacing, usually 16" on center (O.C.) or 24" O.C., with the stud size given. The linear feet of like partitions (having the same stud size, height, and spacing) divided by the spacing will give the estimator the approximate number of studs required. Additional studs for openings, corners, double top plates, sole plates, and intersecting partitions must be taken off separately. An allowance for waste should be included (or heights should be recorded as a standard length). One "rule of thumb" is to allow one stud for every linear foot of wall, for 16" O.C. spacing.

Number and Size of Openings are important. Even though there are no studs in these areas, the estimator must take off headers, subsills, king studs, trimmers, cripples, and knee studs. Where bracing and fire blocking are noted, indicate the type and quantity.

Tongue and Groove Decks of various woods, solid planks, or laminated construction are nominally 2" to 4" thick and are often used with glued laminated beams or heavy timber framing. The square foot method is used to determine quantities and consideration given to non-modular areas for the amount of waste involved. The materials are purchased by board foot measurement. The conversion from square foot to board foot must allow for net sizes as opposed to board measure. In this way, loss of coverage due to the tongue and available mill lengths can be taken into account.

Sheathing on Walls can be plywood of different grades and thicknesses, wallboard, or solid boards nailed directly to the studs. Plywood may be applied with the grain vertical, horizontal, or rarely, diagonal to the studding. Solid boards are usually nailed diagonally, but can be applied horizontally when lateral forces are not present. For solid board sheathing, add 15% to 20% more material to the takeoff when using tongue and groove (as opposed to square edge) sheathing. Wallboard can be installed either horizontally or vertically, depending upon wall height and fire code restrictions. When estimating quantities of plywood or wallboard sheathing, the estimator calculates the number of sheets required by measuring the square feet of area to be covered, adding waste, and then dividing by sheet size. Applying these materials diagonally or on non-modular areas will create additional waste. This waste factor must be included in the estimate. For diagonal application of boards, plywood, or wallboard, include an additional 10% to 15% material waste factor.

Subfloors can be CDX type plywood (with the thickness dependent on the load and span), solid boards laid diagonally or perpendicular to the joists, or tongue and groove planks. The quantity takeoff for subfloors is similar to sheathing (noted above).

Grounds are normally 1" x 2" wood strips used for case work or plaster; the quantities are estimated in L.F.

Furring (1" x 2" or 3") wood strips are fastened to wood, masonry or concrete walls so that wall coverings may be attached thereto. Furring may also be used on the underside of ceiling joists to fasten ceiling finishes. Quantities are estimated by L.F.

The linear feet of framing members (studs or joists) required is based on square feet of surface area (wall, floor, ceiling).

Spacing of Framing Members	Linear Feet per Square Foot Surface
12" O.C.	1.2 L.F./S.F.
16" O.C.	1.0 L.F./S.F.
24" O.C.	0.8 L.F./S.F.

The requirements for rough carpentry, especially those for temporary construction, may not all be directly stated in the plans and specifications. These additional items may include blocking, temporary stairs, wood inserts for metal pan stairs, and railings, along with various other requirements for different trades. Temporary construction may also be included in Division 1 – General Requirements.

Laminated Construction

Laminated construction should be listed separately, as it is frequently supplied by a specialty subcontractor. Sometimes the beams are supplied and erected by one subcontractor, and the decking installed by the general contractor or another subcontractor. The takeoff units must be adapted to the system: square foot (floor), linear foot (members), or board foot (lumber). Since the members are factory fabricated, the plans and specifications must be submitted to a fabricator for takeoff and pricing. Examples of rough carpentry and laminated construction are shown in Figures 6.1a, 6.1b, 6.1c, and 6.1d.

Finish Carpentry and Millwork

Finish carpentry and millwork – wood rails, paneling, shelves, casements, and cabinetry – are common in buildings that have no other wood. Upon examination of the plans and specifications, the estimator must determine which items will be built on-site, and which will be fabricated off-site by a millwork subcontractor. Shop drawings are often required for architectural woodwork and are usually included in the subcontract price.

Mouldings and Door Trim may be taken off and priced by the "set" or by the linear foot. The common use of pre-hung doors makes it convenient to take off this trim with the doors. Exterior trim, other than door and window trim, should be taken off with the siding, since the details and dimensions are interrelated.

Paneling is taken off by type, finish, and square foot (converted to full sheets). Be sure to list any millwork that would show up on the details. Panel siding and associated trim are taken off by the square foot and linear foot, respectively. Be sure to provide an allowance for waste.

Decorative Beams and Columns that are non-structural should be estimated separately. Decorative trim may be used to wrap exposed structural elements. Particular attention should be paid to the joinery. Long, precise joints are difficult to construct in the field.

Cabinets, Counters and Shelves are most often priced by the linear foot or by the unit. Job fabricated, prefabricated, and subcontracted work should be estimated separately.

Stairs should be estimated by individual component unless accurate, complete system costs have been developed from previous projects. Typical components and units for estimating are shown in Figure 6.2.

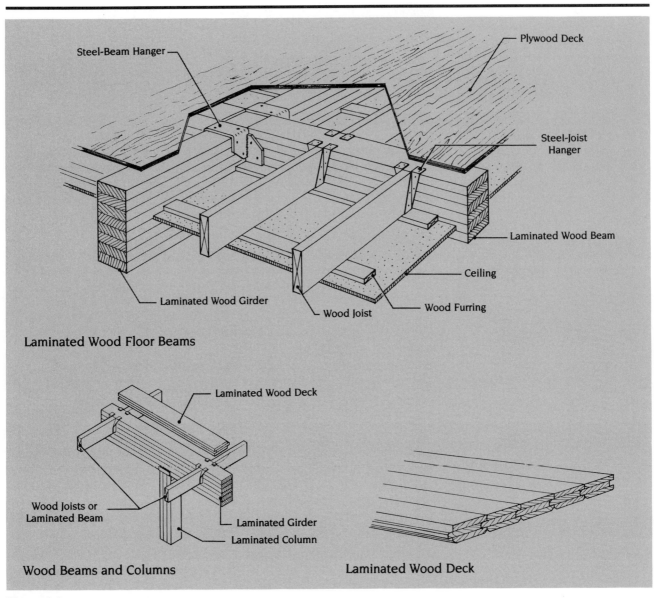

Laminated Wood Floor Beams

Wood Beams and Columns

Laminated Wood Deck

Figure 6.1a

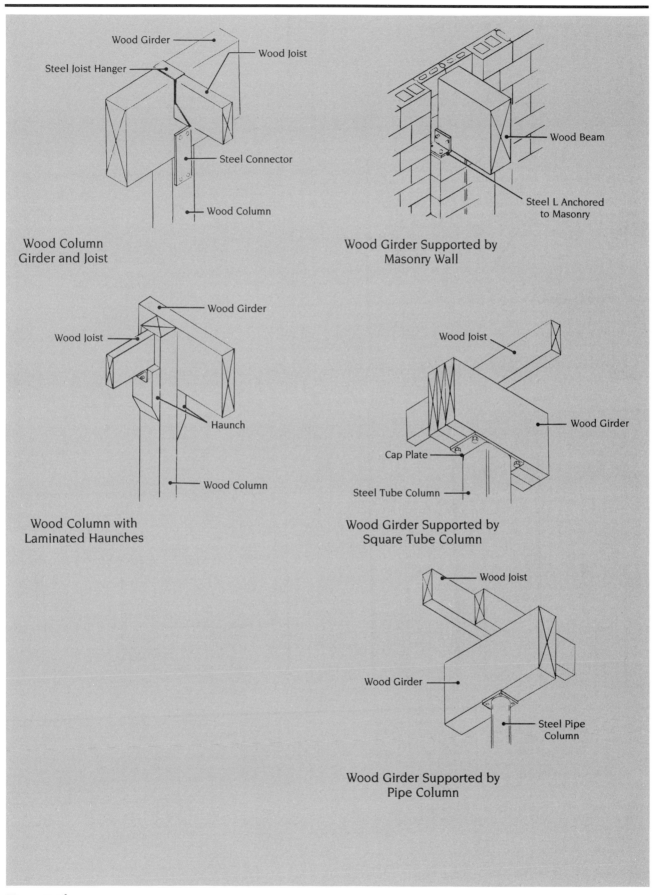

Wood Girder

Wood Joist

Steel Joist Hanger

Steel Connector

Wood Column

**Wood Column
Girder and Joist**

Wood Beam

Steel L Anchored
to Masonry

**Wood Girder Supported by
Masonry Wall**

Wood Girder

Wood Joist

Haunch

Wood Column

**Wood Column with
Laminated Haunches**

Wood Joist

Wood Girder

Cap Plate

Steel Tube Column

**Wood Girder Supported by
Square Tube Column**

Wood Joist

Wood Girder

Steel Pipe
Column

**Wood Girder Supported by
Pipe Column**

Figure 6.1b

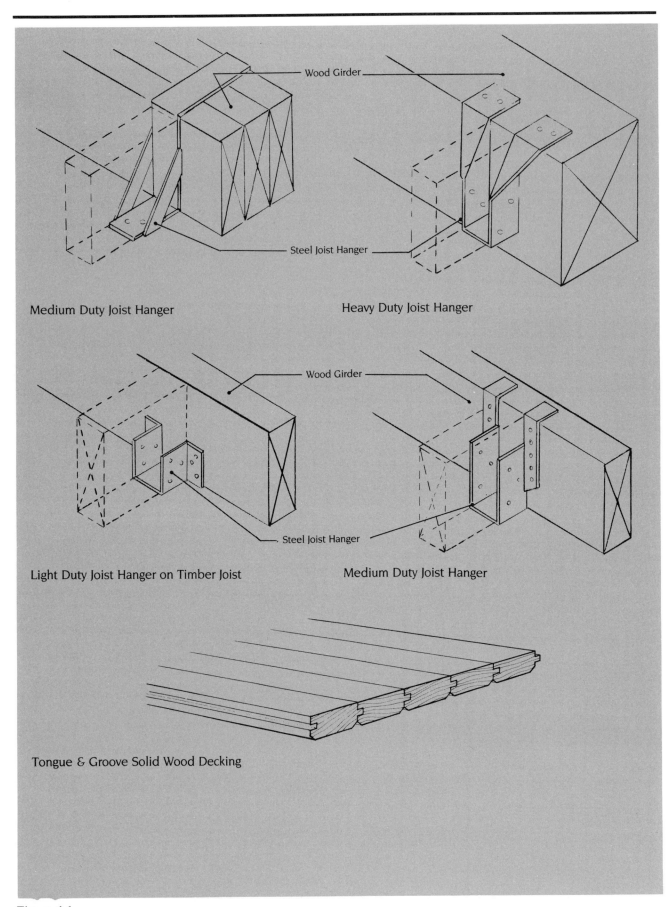

Medium Duty Joist Hanger

Heavy Duty Joist Hanger

Wood Girder

Steel Joist Hanger

Light Duty Joist Hanger on Timber Joist

Wood Girder

Steel Joist Hanger

Medium Duty Joist Hanger

Tongue & Groove Solid Wood Decking

Figure 6.1c

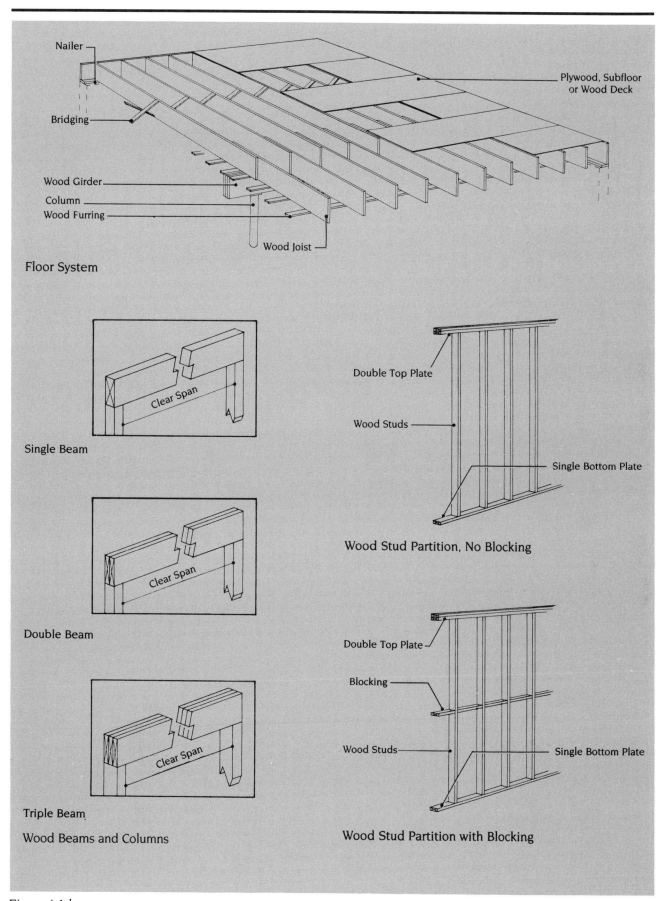

Nailer

Plywood, Subfloor or Wood Deck

Bridging

Wood Girder

Column

Wood Furring

Wood Joist

Floor System

Single Beam

Double Beam

Triple Beam

Clear Span

Wood Beams and Columns

Double Top Plate

Wood Studs

Single Bottom Plate

Wood Stud Partition, No Blocking

Double Top Plate

Blocking

Wood Studs

Single Bottom Plate

Wood Stud Partition with Blocking

Figure 6.1d

		6.2 Finish Carpentry	CREW	DAILY OUTPUT	MAN-HOURS	UNIT	BARE COSTS MAT.	LABOR	EQUIP.	TOTAL	TOTAL INCL O&P	
700	0030	Siding, hardboard, 7/16" thick, prime painted, lap,										700
	0050	plain or grooved finish	F-2	750	.021	S.F.	.46	.44	.03	.93	1.18	
	0100	Board finish, 7/16" thick, lap or grooved, primed		750	.021		.71	.44	.03	1.18	1.45	
	0200	Stained		750	.021		.76	.44	.03	1.23	1.51	
	0700	Particle board, overlaid, 3/8" thick		750	.021		.64	.44	.03	1.11	1.37	
	0900	Plywood, medium density overlaid, 3/8" thick		750	.021		.70	.44	.03	1.17	1.44	
	1000	1/2" thick		700	.023		.78	.47	.03	1.28	1.58	
	1100	3/4" thick		650	.025		.92	.51	.03	1.46	1.79	
	1600	Texture 1-11, cedar, 5/8" thick, natural		675	.024		1.33	.49	.03	1.85	2.21	
	1700	Factory stained		675	.024		1.38	.49	.03	1.90	2.27	
	1900	Texture 1-11, fir, 5/8" thick, natural		675	.024		.49	.49	.03	1.01	1.29	
	2000	Factory stained		675	.024		.54	.49	.03	1.06	1.34	
	2050	Texture 1-11, S.Y.P., 5/8" thick, natural		675	.024		.55	.49	.03	1.07	1.36	
	2100	Factory stained		675	.024		.60	.49	.03	1.12	1.41	
	2200	Rough sawn cedar, 3/8" thick, natural		675	.024		.99	.49	.03	1.51	1.84	
	2300	Factory stained		675	.024		1.04	.49	.03	1.56	1.89	
	2500	Rough sawn fir, 3/8" thick, natural		675	.024		.40	.49	.03	.92	1.19	
	2600	Factory stained		675	.024		.45	.49	.03	.97	1.25	
	2800	Redwood, textured siding, 5/8" thick		675	.024		1.15	.49	.03	1.67	2.02	
	3000	Polyvinyl chloride coated, 3/8" thick		750	.021		.79	.44	.03	1.26	1.54	
730	0010	**SOFFITS** Wood fiber, no vapor barrier, 15/32" thick		525	.030		.42	.63	.04	1.09	1.42	730
	0100	5/8" thick		525	.030		.52	.63	.04	1.19	1.53	
	0300	As above, 5/8" thick, with factory finish		525	.030		.60	.63	.04	1.27	1.62	
	0500	Hardboard, 3/8" thick, slotted		525	.030		.64	.63	.04	1.31	1.66	
	1000	Exterior AC plywood, 1/4" thick		420	.038		.40	.78	.05	1.23	1.64	
	1100	1/2" thick		420	.038		.55	.78	.05	1.38	1.81	
760	0010	**STAIR PARTS** Balusters, turned, 30" high, pine, minimum (84)	1 Carp	28	.286	Ea.	3.73	5.85		9.58	12.70	760
	0100	Maximum		26	.308		5	6.30		11.30	14.80	
	0300	30" high birch balusters, minimum		28	.286		4.75	5.85		10.60	13.85	
	0400	Maximum		26	.308		6	6.30		12.30	15.90	
	0600	42" high, pine balusters, minimum		27	.296		4.40	6.10		10.50	13.80	
	0700	Maximum		25	.320		5.50	6.60		12.10	15.70	
	0900	42" high birch balusters, minimum		27	.296		6	6.10		12.10	15.55	
	1000	Maximum		25	.320		6.80	6.60		13.40	17.15	
	1050	Baluster, stock pine, 1-1/16" x 1-1/16"		240	.033	L.F.	.46	.69		1.15	1.51	
	1100	1-5/8" x 1-5/8"		220	.036	"	.93	.75		1.68	2.12	
	1200	Newels, 3-1/4" wide, starting, minimum		7	1.140	Ea.	25	23		48	62	
	1300	Maximum		6	1.330		200	27		227	260	
	1500	Landing, minimum		5	1.600		35	33		68	87	
	1600	Maximum		4	2		210	41		251	290	
	1800	Railings, oak, built-up, minimum		60	.133	L.F.	3.75	2.74		6.49	8.15	
	1900	Maximum		55	.145		8	2.99		10.99	13.20	
	2100	Add for sub rail		110	.073		1.70	1.49		3.19	4.06	
	2110											
	2300	Risers, Beech, 3/4" x 7-1/2" high	1 Carp	64	.125	L.F.	3.95	2.57		6.52	8.10	
	2400	Fir, 3/4" x 7-1/2" high		64	.125		1.05	2.57		3.62	4.93	
	2600	Oak, 3/4" x 7-1/2" high		64	.125		3.15	2.57		5.72	7.25	
	2800	Pine, 3/4" x 7-1/2" high		66	.121		1.05	2.49		3.54	4.81	
	2850	Skirt board, pine, 1" x 10"		55	.145		1.36	2.99		4.35	5.90	
	2900	1" x 12"		52	.154		1.70	3.16		4.86	6.50	
	3000	Treads, oak, 1-1/16" x 9-1/2" wide, 3' long		18	.444	Ea.	14.15	9.15		23.30	29	
	3100	4' long		17	.471		19.80	9.65		29.45	36	
	3300	1-1/16" x 11-1/2" wide, 3' long		18	.444		17	9.15		26.15	32	
	3400	6' long		14	.571		36	11.75		47.75	57	
	3600	Beech treads, add					40%					
	3800	For mitered return nosings, add				L.F.	1.83			1.83	2.01	
790	0011	**STAIRS, PREFABRICATED**										790
	0100	Box stairs, prefabricated, 3'-0" wide										
	0110	Oak treads, no handrails, 2' high	2 Carp	5	3.200	Flight	85	66		151	190	
	0200	4' high	"	4	4	"	200	82		282	340	

60

Figure 6.2

A general rule for budgeting millwork is that total costs will be two to three times the cost of the materials. Millwork is often ordered and purchased directly by the owner. When installation is the responsibility of the contractor, costs for handling, storage, and protection, as well as those for installation, should be included. Typical finish carpentry and millwork items are illustrated in Figure 6.3a and 6.3b.

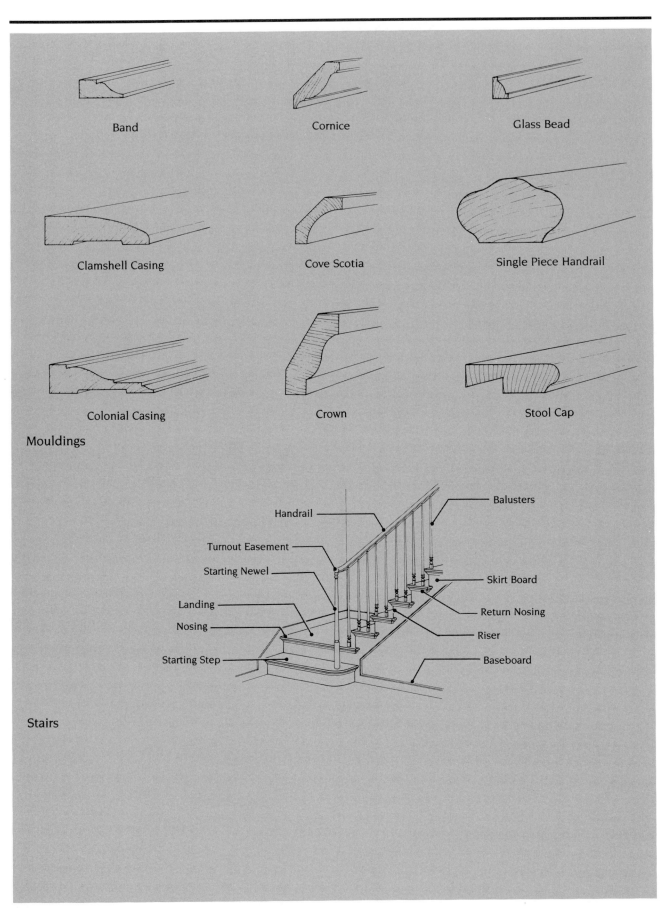

Band

Cornice

Glass Bead

Clamshell Casing

Cove Scotia

Single Piece Handrail

Colonial Casing

Crown

Stool Cap

Mouldings

Handrail

Balusters

Turnout Easement

Starting Newel

Skirt Board

Landing

Return Nosing

Nosing

Riser

Starting Step

Baseboard

Stairs

Figure 6.3a

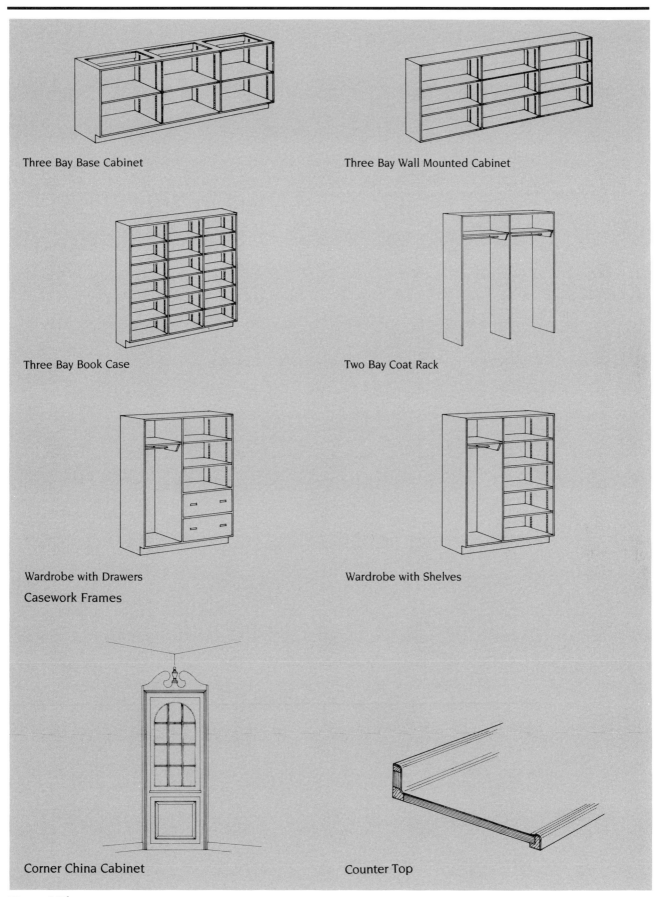

Three Bay Base Cabinet

Three Bay Wall Mounted Cabinet

Three Bay Book Case

Two Bay Coat Rack

Wardrobe with Drawers

Wardrobe with Shelves

Casework Frames

Corner China Cabinet

Counter Top

Figure 6.3b

57

Chapter 7
MOISTURE-THERMAL CONTROL

Items normally included in Division 7 of the CSI Masterformat, Moisture-Thermal Control, are those construction items which protect the interior space of a building from the elements. Primarily, the division includes:

- Roofing
- Siding
- Sheet Metal
- Waterproofing
- Insulation
- Roof Accessories

Most of this type of work would not be part of the interior construction of a building, but certain components are often used which, while easily overlooked, must be included in the estimate.

Caulking and sealants are often considered incidental, however, costs for caulking can be significant depending on the type used and the size and length of the joint. While used for cosmetic purposes before painting, acoustical applications for caulking are also common. If every interior partition must be caulked at the floor and ceiling, material and labor costs can be substantial.

Insulation is used for thermal purposes in the renovation of existing spaces when exterior wall cavities or roof areas are opened. Often the choice of insulation type is based not only on cost, but also on R-value (resistance to heat flow) and thickness. Figure 7.1, from Means *Interior Cost Data*, 1987, illustrates how these variables can be compared to make the best choice. For interiors, batt insulation is often used in the stud of partitions to reduce sound transmission.

Roof Accessories include access latches, smoke hatches, skylights, and sky roofs. Costs for budget purposes are included in Means *Interior Cost Data*, 1987. But because of the specialized nature of these products, prices should be obtained from vendors or subcontractors. Examples are illustrated in Figure 7.2.

			DAILY	MAN-		BARE COSTS				TOTAL	
7.2 Insulation		CREW	OUTPUT	HOURS	UNIT	MAT.	LABOR	EQUIP.	TOTAL	INCL O&P	
800	0060	1-½" thick, R6.2	1 Carp	1,000	.008	S.F.	.35	.16		.51	.63	800
	0080	2" thick, R8.3		1,000	.008		.46	.16		.62	.75	
	0120	3" thick, R12.4		800	.010		.70	.21		.91	1.07	
	0370	3#/C.F., unfaced, 1" thick, R4.3		1,000	.008		.50	.16		.66	.79	
	0390	1-½" thick, R6.5		1,000	.008		.75	.16		.91	1.07	
	0400	2" thick, R8.7		890	.009		1.01	.18		1.19	1.38	
	0420	2-½" thick, R10.9		800	.010		1.26	.21		1.47	1.69	
	0440	3" thick, R13		800	.010		1.50	.21		1.71	1.95	
	0520	Foil faced, 1" thick, R4.3		1,000	.008		.94	.16		1.10	1.28	
	0540	1-½" thick, R6.5		1,000	.008		1.18	.16		1.34	1.54	
	0560	2" thick, R8.7		890	.009		1.43	.18		1.61	1.84	
	0580	2-½" thick, R10.9		800	.010		1.68	.21		1.89	2.15	
	0600	3" thick, R13		800	.010		2.73	.21		2.94	3.30	
	0670	6#/C.F., unfaced, 1" thick, R4.3		1,000	.008		.91	.16		1.07	1.24	
	0690	1-½" thick, R6.5		890	.009		1.37	.18		1.55	1.78	
	0700	2" thick, R8.7		800	.010		1.82	.21		2.03	2.30	
	0721	2-½" thick, R10.9		800	.010		2.28	.21		2.49	2.81	
	0741	3" thick, R13		730	.011		2.75	.23		2.98	3.36	
	0821	Foil faced, 1" thick, R4.3		1,000	.008		1.32	.16		1.48	1.69	
	0840	1-½" thick, R6.5		890	.009		1.76	.18		1.94	2.21	
	0850	2" thick, R8.7		800	.010		2.21	.21		2.42	2.73	
	0880	2-½" thick, R10.9		800	.010		2.63	.21		2.84	3.19	
	0900	3" thick, R13		730	.011		3.10	.23		3.33	3.74	
	1500	Foamglass, 1-½" thick, R2.64		800	.010		1.42	.21		1.63	1.86	
	1550	2" thick, R5.26		730	.011		1.98	.23		2.21	2.51	
	1700	Perlite, 1" thick, R2.77		800	.010		.28	.21		.49	.61	
	1750	2" thick, R5.55		730	.011		.56	.23		.79	.95	
	1900	Polystyrene, extruded blue, 2.2#/C.F., ¾" thick, R4		800	.010		.36	.21		.57	.70	
	1940	1-½" thick, R8.1		730	.011		.59	.23		.82	.98	
	1960	2" thick, R10.8		730	.011		.78	.23		1.01	1.19	
	2100	Molded bead board, white, 1" thick, R3.85		800	.010		.15	.21		.36	.47	
	2120	1-½" thick, R5.6		730	.011		.23	.23		.46	.58	
	2140	2" thick, R7.7		730	.011		.31	.23		.54	.67	
	2350	Sheathing, insulating foil faced fiberboard, ⅜" thick		670	.012		.21	.25		.46	.59	
	2510	Urethane, no paper backing, ½" thick, R2.9		800	.010		.22	.21		.43	.54	
	2520	1" thick, R5.8		800	.010		.42	.21		.63	.76	
	2540	1-½" thick, R8.7		730	.011		.63	.23		.86	1.02	
	2560	2" thick, R11.7		730	.011		.84	.23		1.07	1.25	
	2710	Fire resistant, ½" thick, R2.9		800	.010		.27	.21		.48	.60	
	2720	1" thick, R5.8		800	.010		.57	.21		.78	.93	
	2740	1-½" thick, R8.7		730	.011		.78	.23		1.01	1.19	
	2760	2" thick, R11.7		730	.011		1.04	.23		1.27	1.47	
850	0010	**WALL OR CEILING INSULATION, NON-RIGID**										850
	0040	Fiberglass, kraft faced, batts or blankets										
	0060	3-½" thick, R11, 11" wide	1 Carp	1,150	.007	S.F.	.20	.14		.34	.43	
	0080	15" wide		1,600	.005		.20	.10		.30	.37	
	0100	23" wide		1,600	.005		.22	.10		.32	.39	
	0140	6" thick, R19, 11" wide		1,000	.008		.32	.16		.48	.59	
	0160	15" wide		1,350	.006		.32	.12		.44	.53	
	0180	23" wide		1,600	.005		.32	.10		.42	.50	
	0200	9" thick, R30, 15" wide		1,150	.007		.50	.14		.64	.76	
	0220	23" wide		1,350	.006		.50	.12		.62	.73	
	0240	12" thick, R38, 15" wide		1,000	.008		.67	.16		.83	.98	
	0260	23" wide		1,350	.006		.67	.12		.79	.92	
	0400	Fiberglass, foil faced, batts or blankets										
	0420	3-½" thick, R11, 15" wide	1 Carp	1,600	.005	S.F.	.22	.10		.32	.39	
	0440	23" wide		1,600	.005		.22	.10		.32	.39	
	0460	6" thick, R19, 15" wide		1,350	.006		.34	.12		.46	.55	
	0480	23" wide		1,600	.005		.34	.10		.44	.52	
	0500	9" thick, R30, 15" wide		1,150	.007		.53	.14		.67	.79	

69

Figure 7.1

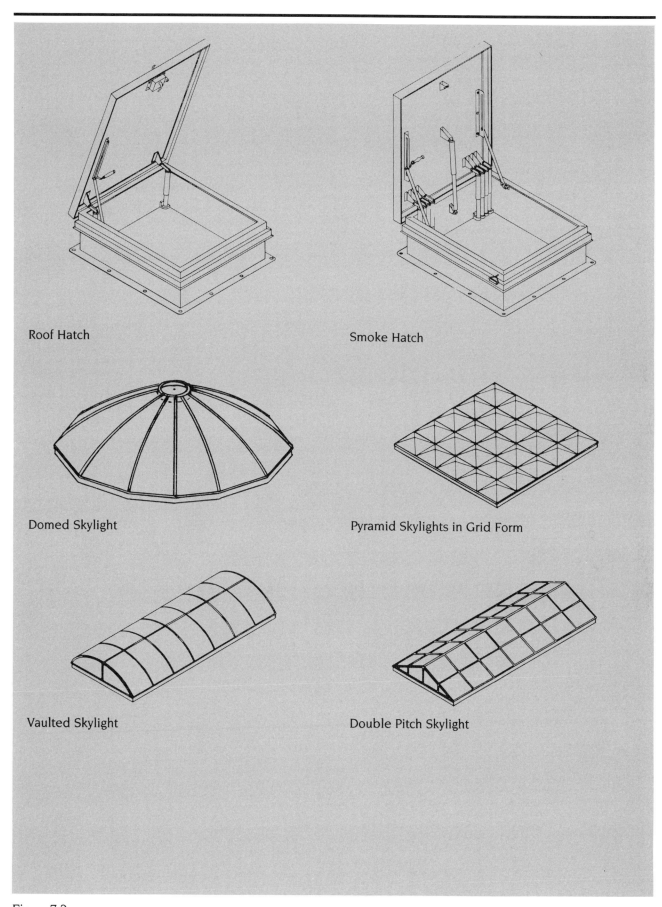

Roof Hatch

Smoke Hatch

Domed Skylight

Pyramid Skylights in Grid Form

Vaulted Skylight

Double Pitch Skylight

Figure 7.2

Chapter 8
DOORS, WINDOWS AND GLASS

The cost of doors, windows, and glass, when part of the interior construction contract, can represent a sizeable portion of the total cost. Windows and doors require hardware, an additional material cost. Any one door assembly (door, frame, hardware) can be a combination of many variable features, shown below:

Door:	Frame:	Hardware:
Size	Size	Lockset
Thickness	Throat	Passage set
Wood-type	Wood-type	Panic bar
Metal-gauge	Metal gauge	Closer
Laminate	Casing	Hinges
Hollow-core type	Stops	Stops
Solid-core material	Fire rating	Bolts
Fire rating	Knock down	Finish
Finish	Welded	Plates

Most architectural plans and specifications include door, window, and hardware schedules to tabulate these combinations. The estimator should use these schedules and details in conjunction with the plans to avoid duplication or omission of units when determining the quantities. The schedules should identify the location, size, and type of each unit. Schedules should also include information regarding the frame, fire-rating, hardware, and special notes. If no such schedules are included, the estimator should prepare them in order to provide an accurate quantity takeoff. Figure 8.1 is an example of a schedule that may be prepared by the estimator. Most suppliers will prepare separate schedules; each must be approved by the architect or owner.

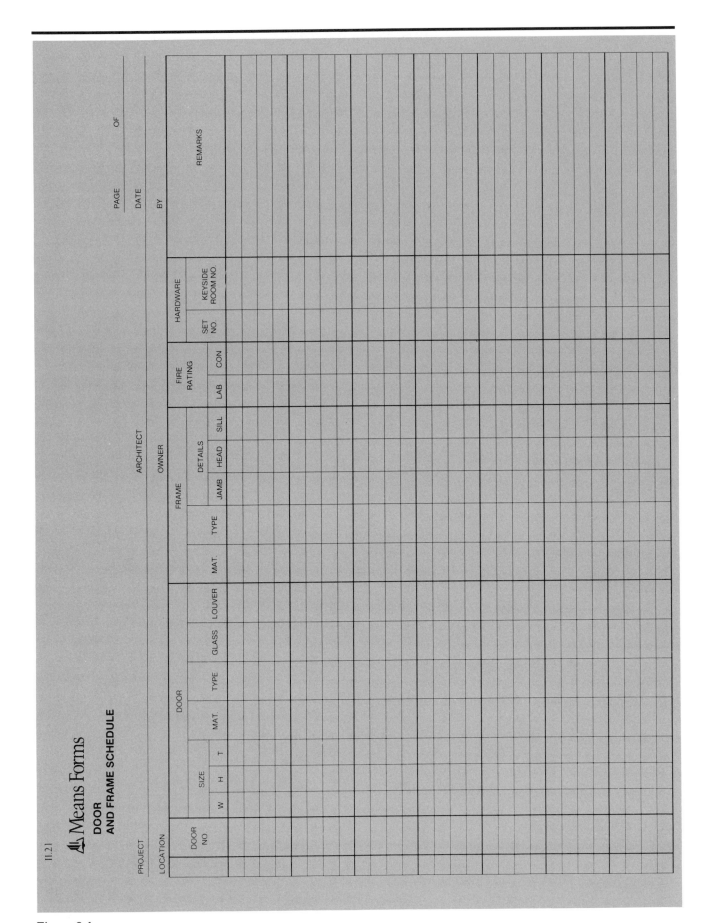

Figure 8.1

64

A proper door schedule on the architectural drawings identifies each opening in detail. Define each opening in accordance with the items in the schedule and any other pertinent data. Installation information should be carefully reviewed in the specifications.

For the quantity takeoff, the numbers of all similar doors and frames should be combined. Each should be checked off on the plans to ensure none have been left out. An easy and obvious double-check is to count the total number of openings, making certain that two doors and only one frame have been included where double doors are used. Important details to check for both door and frame are the following:

- Material
- Gauge
- Size
- Core Material
- Fire Rating Label
- Finish
- Style

Wood Doors

Wood doors are manufactured in either flush or paneled designs, and are separated into three grades; architectural/commercial, residential, and decorator. A wide variety of frames are available for interior installations in metal, pine, hardwood, and for various partition thicknesses. Some doors are available pre-hung for quick installation.

Architectural or commercial wood doors are the type most often specified in building construction. The stiles are made of hardwood, and the core is dense and of hot-bonded construction. They feature thick face veneers that are exterior-glued and matched in their grain patterns. Because of its durability, this grade of door often carries a lifetime warranty.

Residential wood doors are used in low-use situations where economy is a primary consideration. The stiles are manufactured from soft wood; the core from low density materials. The face veneers are thin, interior-glued, and broken in their grain patterns.

Decorator wood doors are manufactured from solid wood and are usually hand carved. Because of the choice woods used and the special craftsmanship required in their production, their cost is several times that of similar size architectural wood doors.

Flush doors are produced with cores of varying density: hollow, particle-board, or solid wood block. Lauan mahogany, birch, or other hardwood veneers are used in their facings. Synthetic veneers, created from a medium-density overlay or high-pressure plastic laminate, may serve as an alternate choice to natural wood veneers. Flush wood doors may be fire rated. Door swings and wood door details are shown in Figure 8.2.

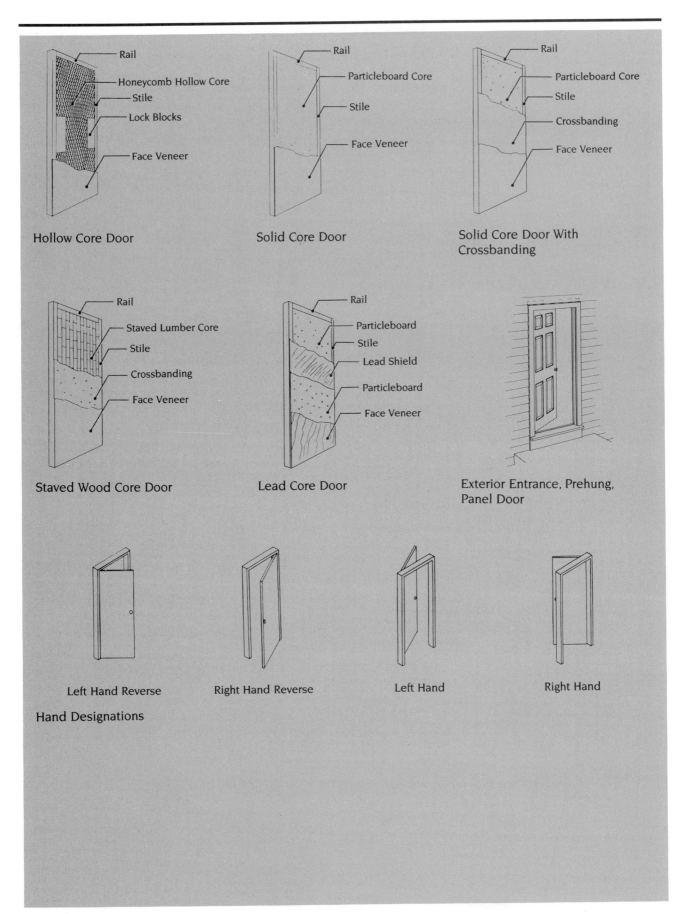

Hollow Core Door
- Rail
- Honeycomb Hollow Core
- Stile
- Lock Blocks
- Face Veneer

Solid Core Door
- Rail
- Particleboard Core
- Stile
- Face Veneer

Solid Core Door With Crossbanding
- Rail
- Particleboard Core
- Stile
- Crossbanding
- Face Veneer

Staved Wood Core Door
- Rail
- Staved Lumber Core
- Stile
- Crossbanding
- Face Veneer

Lead Core Door
- Rail
- Particleboard
- Stile
- Lead Shield
- Particleboard
- Face Veneer

Exterior Entrance, Prehung, Panel Door

Hand Designations
- Left Hand Reverse
- Right Hand Reverse
- Left Hand
- Right Hand

Figure 8.2

The estimator must pay particular attention to fire door specifications when performing the quantity takeoff. It is important to determine the exact type of door required. Figure 8.3 is a table describing various types of fire doors. Please note that a "B" label door can be one of four types. If the plans or door schedule do not specify exactly which temperature rise is required, the estimator should consult the architect or local building inspector. Many building and fire codes also require that frames and hardware at fire doors be fire rated and labeled as such. When determining quantities, the estimator must also include any glass (usually wired) or special inserts to be installed in fire doors (or in any doors).

Fire Door			
Classification	Time Rating (as Shown on Label)	Temperature Rise (as Shown on Label)	Maximum Glass Area
3 Hour fire doors (A) are for use in openings in walls separating buildings or dividing a single building into the areas.	3 Hr. (A) 3 Hr. (A) 3 Hr. (A) 3 Hr. (A)	30 Min. 250°F Max 30 Min. 450°F Max 30 Min. 650°F Max *	None
1-1/2 Hour fire doors (B) and (D) are for use in openings in 2 Hour enclosures of vertical communication through buildings (stairs, elevators, etc.) or in exterior walls which are subject to severe fire exposure from outside of the building. 1 Hour fire doors (B) are for use in openings in 1 Hour enclosures of vertical communication through buildings (stairs, elevators, etc.)	1-1/2 Hr. (B) 1-1/2 Hr. (B) 1-1/2 Hr. (B) 1-1/2 Hr. (B) 1 Hr. 1-1/2 Hr. (D) 1-1/2 Hr. (D) 1-1/2 Hr. (D) 1-1/2 Hr. (D)	30 Min. 250°F Max 30 Min. 450°F Max 30 Min. 650°F Max * 30 Min. 250°F Max 30 Min. 250°F Max 30 Min. 450°F Max 30 Min. 650°F Max *	100 square inches per door None
3/4 Hour fire doors (C) and (E) are for use in openings in corridor and room partitions or in exterior walls which are subject to moderate fire exposure from outside of the building.	3/4 Hr. (C) 3/4 Hr. (E)	** **	1296 Square 720 square inches per light
1/2 Hour fire doors and 1/3 Hour fire doors are for use where smoke controls is a primary consideration and are for the protection of openings in partitions between a habitable room and a corridor when the wall has a fire-resistance rating of not more than one hour.	1/2 Hr. 1/3 Hr.	** **	No limit

*The labels do not record any temperature rise limits. This means that the temperature rise on the unexposed face of the door at the end of 30 minutes of test is in excess of 650°F.
**Temperature rise is not recorded.

Figure 8.3

Metal Doors and Frames

Hollow metal doors are available in stock or custom fabrication, flush or embossed, with glazing or louvres, labeled or unlabeled, and in various steel gauges and core fills. Stock doors may be supplied for low-, moderate-, or high-frequency of use from some manufacturers. The doors are available in widths of 2' to 4' and heights varying from 6'-8" to 10' or more. They may be used as single doors, in pairs with both leaves active, in pairs with one active leaf, including an astragal. Bi-fold hollow metal doors are available for shielded applications.

Hollow metal doors are reinforced at the stress points and pre-mortised for the hardware required for the door application. Hollow metal, labelled fire doors can be supplied stock or custom manufactured with A, B, C, D, E labels, with 3/4 to 3 hour ratings, depending on the glass area, height and width restrictions, and maximum expected temperature rise (shown in Figure 8.3). The door types shown in Figure 8.4 are examples of typical metal fire doors. Code requirements for fire doors and ratings vary from state to state and often from city to city.

Hollow metal frames may be supplied in 14, 16, or 18 gauge galvanized or plain steel in knockdown standard frames or welded customized frames that can be fabricated to satisfy most design conditions. Frames with borrowed lights, transoms, or cased openings are available in stick components from some manufacturers. Frames may be wraparound (enclosing the wall) or butt up against the opening. A wraparound frame may terminate into the enclosed wall when it is covered by a finish such as plaster, or the frame may return along the enclosed wall when it is exposed, as in drywall construction. They are sometimes supplied in two pieces to suit varied wall thicknesses, or in one piece to satisfy standard wall thicknesses. Frames are normally reinforced at stress points and are prepared for hinges and strikes. Anchors to attach the frame to the wall are supplied to suit wall construction requirements. Custom frames normally require a hardware schedule and templates to produce required shop drawings and to accomplish fabrication. Typical hollow metal frames for various applications are shown in Figure 8.5.

Special Doors

Metal access panels and doors are available in steel or stainless steel for fire-rated or non-fire-rated applications. Panels are fabricated for flush installations in drywall, both skim coated or taped, for masonry and tile applications, for plastered walls and ceilings, and for acoustical ceilings. The doors are available in stock sizes and types to suit most applications.

Blast doors are available in standard designs. They may be custom designed to withstand specified pressures and to resist penetration. Cold-storage doors are available in standard designs in wood, steel, fiberglass, plastic, and stainless steel for all types of cold storage requirements. These doors are manufactured to provide insulation requirements for cool zones, coolers, and freezers for manual, air, electric, or hydraulic operations. Door operation types include sliding, vertical lift, bi-parting overhead, and single- and double-swing.

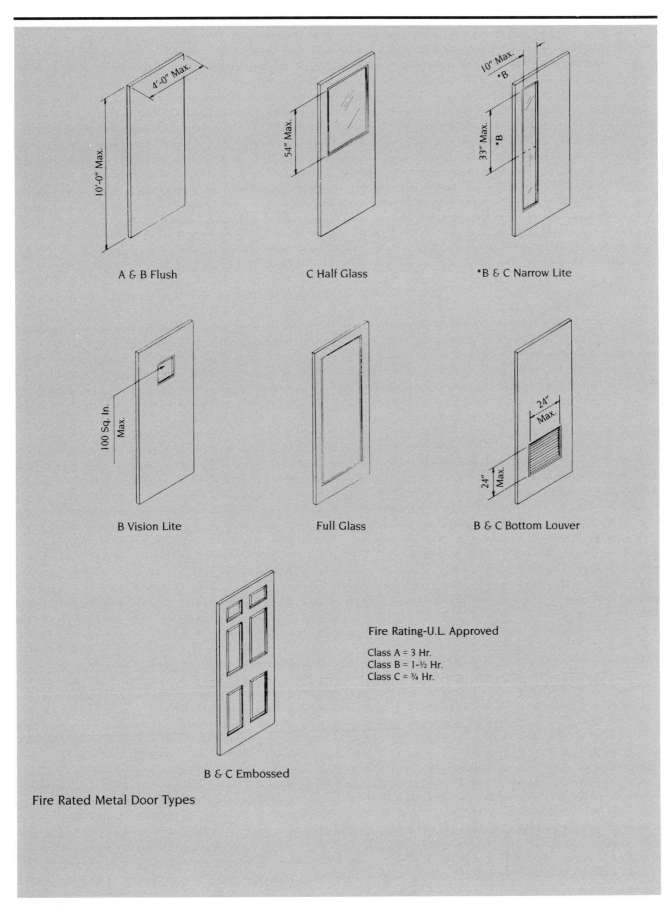

A & B Flush

C Half Glass

*B & C Narrow Lite

B Vision Lite

Full Glass

B & C Bottom Louver

B & C Embossed

Fire Rating-U.L. Approved

Class A = 3 Hr.
Class B = 1-½ Hr.
Class C = ¾ Hr.

Fire Rated Metal Door Types

Figure 8.4

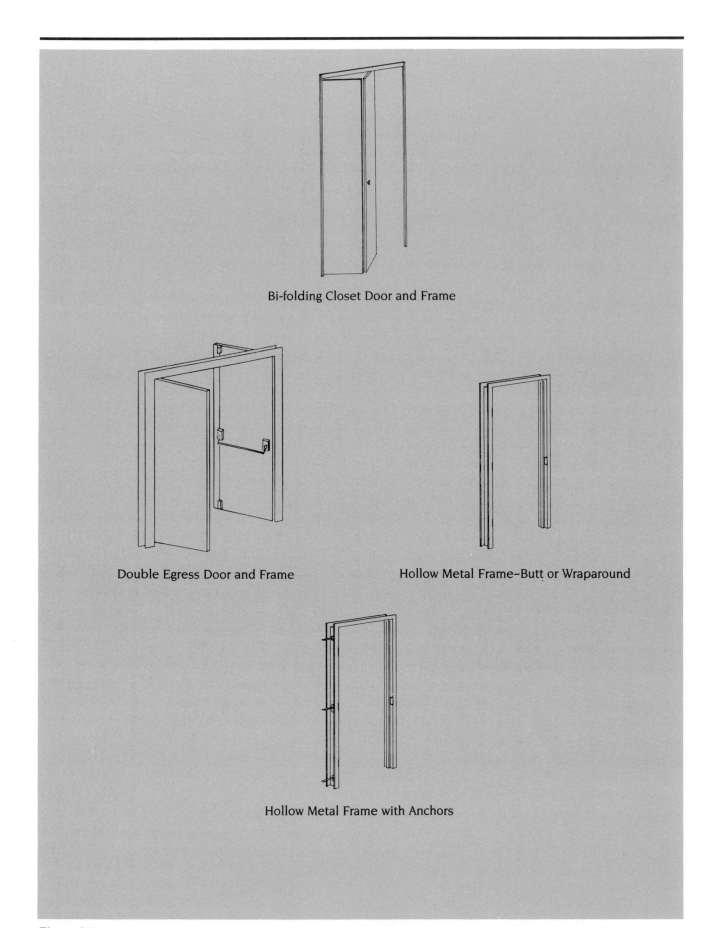

Bi-folding Closet Door and Frame

Double Egress Door and Frame

Hollow Metal Frame–Butt or Wraparound

Hollow Metal Frame with Anchors

Figure 8.5

Pass windows (vertical and horizontal sliding), rotating shelf windows, ticket windows, cashier doors, and coin and cash trays are available in aluminum, steel, or stainless steel. Pass windows, roll up shutters, and projection booth shutters are available, labelled or non-labelled, with fusible links to suit most applications and requirements. A roll-up shutter and a roll up gate are shown in Figure 8.6.

Darkroom revolving doors and pass-throughs are modeled to fit an existing door frame. They provide protection for multiple darkroom installations.

Glass and Glazing

Glass and glazing in interior construction may involve interior glazed partitions, entrances, and storefronts. Interior glazed partitions are commonly constructed of tubular aluminum framing. Glazing subcontractors may estimate such partitions by measuring and pricing the length of each component of the frame separately. Examples are shown in Figures 8.7 and 8.8. The glass can be plate, tempered, safety, tinted, insulated, or combinations thereof, depending on project and code requirements. Glass is estimated by the square foot or by united length (length plus width). Glass doors and hardware may also be estimated separately. Another method of estimating is to use complete system prices, by square foot (Figure 8.9) or by opening (Figure 8.10)

Entrances and storefronts are almost all special designs and combinations of unit items to fit a unique situation. The estimator should submit the plans and specifications to a specialty installer for takeoff and pricing. Typical entrances are shown in Figure 8.11. The general procedure for the installer's takeoff is:

For stationary units:
- Determine height and width of each like unit.
- Determine linear feet of intermediate, horizontal, and vertical members, rounded to next higher foot.
- Determine number of joints.

For entrance units:
- Determine number of joints.
- Determine special frame hardware per unit.
- Determine special door hardware per unit.
- Determine thresholds and closers.

Curtain Walls

A curtain wall is a non-structural facade that may be used as an interior wall. It consists of panels in a wide variety of materials and constructions. A curtain wall is held in place in a metal frame by caulking, gaskets, and sealants. The extent to which the curtain wall can be prefabricated varies by types: *stick*, in which all components are field assembled; *panel and mullion*, in which the panels are prefabricated into frames and field connected to mullions; and *total panel systems*, in which the mullions are pre-assembled into the panels.

The mullions are extruded or fabricated in "I" or tubular sections which supply the strength for lateral loads. The mullions may be one- or two-piece sections with "pockets" to receive cover plate assemblies, gaskets and accent strips, and flanges or recesses to receive panel frames and window sashes. Aluminum is an ideal material for complex sections, due to its extrudability, but bronze and steel (stainless or weathering) can be rolled into less complicated sections. Figure 8.12 shows different types of curtain wall systems.

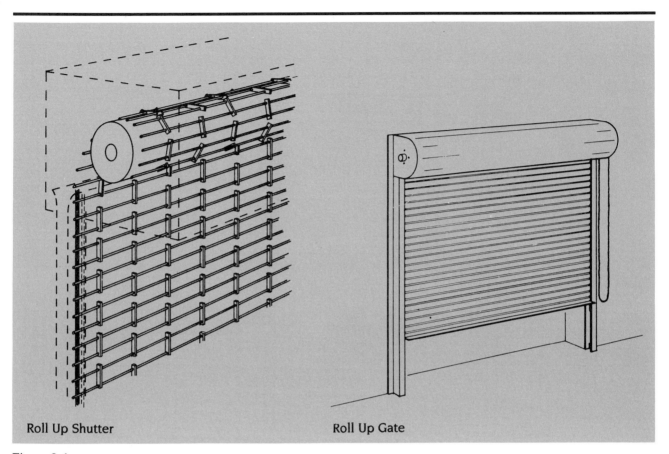

Roll Up Shutter   Roll Up Gate

Figure 8.6

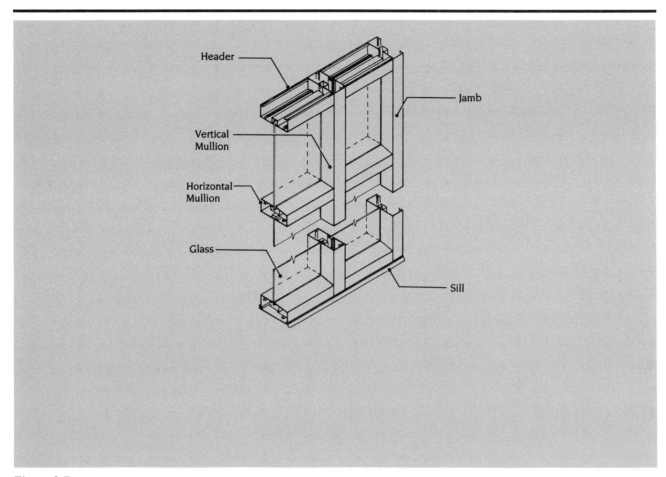

Figure 8.7

8.8 Glass & Glazing

			CREW	DAILY OUTPUT	MAN-HOURS	UNIT	MAT.	LABOR	EQUIP.	TOTAL	TOTAL INCL O&P	
450	0010	REFLECTIVE GLASS ¼" float with fused metallic oxide, tinted	2 Glaz	115	.139	S.F.	5.70	2.89		8.59	10.45	450
	0500	¼" float glass with reflective applied coating		115	.139		3.40	2.89		6.29	7.90	
	2000	Solar film on glass, not including glass, minimum		180	.089		1	1.84		2.84	3.76	
	2050	Maximum		225	.071		1.95	1.48		3.43	4.27	
480	0010	SANDBLASTED GLASS Float glass, ⅛" thick		160	.100		3	2.08		5.08	6.30	480
	0100	3/16" thick		130	.123		3.50	2.55		6.05	7.55	
	0500	Plate glass, ¼" thick		120	.133		3.70	2.77		6.47	8.05	
	0600	⅜" thick		75	.213		4.20	4.43		8.63	11	
540	0010	SPANDREL GLASS ¼" thick, standard colors, over 2000 S.F.		110	.145		4	3.02		7.02	8.75	540
	0200	Under 2000 S.F.		120	.133		5.30	2.77		8.07	9.80	
	0300	For custom colors, add				Total	10%			10%	10%	
	0500	For ⅜" thick, add				S.F.	3.80			3.80	4.18	
	1000	For double coated, ¼" thick, add					1.35			1.35	1.49	
	1200	For insulation on panels, add					2.75			2.75	3.03	
	2000	Panels, insulated , with aluminum backed fiberglass, 1" thick	2 Glaz	120	.133		6.65	2.77		9.42	11.30	
	2100	2" thick	"	120	.133		7.35	2.77		10.12	12.05	
	2500	With galvanized steel backing, add					2.10			2.10	2.31	
	3000	Maximum size 72" x 168", for over 140", add					10%					
600	0010	WINDOW GLASS Clear float, stops, putty bed, ⅛" thick	2 Glaz	480	.033		2.25	.69		2.94	3.47	600
	0500	3/16" thick, clear		480	.033		2.35	.69		3.04	3.58	
	0600	Tinted		480	.033		2.65	.69		3.34	3.91	
	0700	Tempered		480	.033		4.50	.69		5.19	5.95	
630	0010	WINDOW WALL See division 8.9										
660	0010	WIRE GLASS ¼" thick, rough obscure (chicken wire)	2 Glaz	135	.119	S.F.	4.60	2.46		7.06	8.60	660
	1000	Polished wire, ¼" thick, diamond, clear		135	.119		5.20	2.46		7.66	9.25	
	1500	Pinstripe, obscure		135	.119		7.85	2.46		10.31	12.20	

8.9 Window/Curtain Walls

			CREW	DAILY OUTPUT	MAN-HOURS	UNIT	MAT.	LABOR	EQUIP.	TOTAL	TOTAL INCL O&P	
100	0010	CURTAIN WALLS Aluminum, stock, including glazing, minimum	H-1	205	.156	S.F.	15	3.34		18.34	22	100
	0050	Average, single glazed		195	.164		20	3.52		23.52	27	
	0150	Average, double glazed		180	.178		35	3.81		38.81	44	
	0200	Maximum		160	.200		52	4.29		56.29	64	
500	0010	TUBE FRAMING For window walls and store fronts, aluminum, stock										500
	0050	Plain tube frame, mill finish, 1-¾" x 1-¾"	2 Glaz	103	.155	L.F.	4.20	3.22		7.42	9.25	
	0100	1-¾" x 3"		100	.160		4.80	3.32		8.12	10.05	
	0150	1-¾" x 4"		98	.163		5.75	3.39		9.14	11.20	
	0200	1-¾" x 4-½"		95	.168		6.10	3.49		9.59	11.75	
	0250	2" x 6"		89	.180		8	3.73		11.73	14.20	
	0300	3" x 3"		92	.174		7.45	3.61		11.06	13.40	
	0350	4" x 4"		87	.184		8.20	3.82		12.02	14.50	
	0400	4-½" x 4-½"		85	.188		10.50	3.91		14.41	17.20	
	0450	Glass bead		240	.067		1.35	1.38		2.73	3.48	
	1000	Flush tube frame, mill finish, ¼" glass, 1-¾" x 4", open header		80	.200		4.20	4.15		8.35	10.60	
	1050	Open sill		82	.195		4.25	4.05		8.30	10.50	
	1100	Closed back header		83	.193		5.60	4		9.60	11.90	
	1150	Closed back sill		85	.188		6	3.91		9.91	12.25	
	1200	Vertical mullion, one piece		75	.213		5.75	4.43		10.18	12.70	
	1250	Two piece		73	.219		6.40	4.55		10.95	13.60	
	1300	90° or 180° vertical corner post		75	.213		10.40	4.43		14.83	17.80	
	1400	1-¾" x 4-½", open header		80	.200		4.85	4.15		9	11.30	
	1450	Open sill		82	.195		5.05	4.05		9.10	11.40	
	1500	Closed back header		83	.193		5.65	4		9.65	12	

Figure 8.8

8.4	Entrances & Storefronts	CREW	DAILY OUTPUT	MAN-HOURS	UNIT	BARE COSTS				TOTAL	
						MAT.	LABOR	EQUIP.	TOTAL	INCL O&P	
100 0500	Stainless steel and glass, 3' x 7', economy	2 Sswk	.90	17.780	Ea.	3,735	395		4,130	4,725	100
0600	Premium	"	.70	22.860	"	6,360	505		6,865	7,800	
200 0010	REVOLVING DOORS 6'-6" to 7'-0" diameter										200
0020	6'-10" to 7' high, stock units, minimum	4 Sswk	.75	42.670	Opng.	16,000	945		16,945	19,100	
0050	Average		.60	53.330		20,500	1,175		21,675	24,400	
0100	Maximum		.45	71.110		25,000	1,575		26,575	30,000	
1000	Stainless steel		.30	107		31,000	2,350		33,350	37,900	
1100	Solid bronze	↓	.15	213		37,000	4,725		41,725	48,200	
1500	For automatic controls, add	2 Elec	2	8	↓	3,200.	180		3,380	3,775	
250 0010	SLIDING ENTRANCE 12' x 7'-6" opng., 5' x 7' door, 2 way traf.,										250
0020	mat activated, panic pushout, incl. operator & hardware,										
0030	not including glass or glazing	2 Glaz	.70	22.860	Opng.	7,400	475		7,875	8,825	
300 0010	SLIDING PANEL Mall fronts, aluminum & glass, 15' x 9' high	2 Glaz	1.30	12.310	Opng.	1,800	255		2,055	2,350	300
0100	24' x 9' high		.70	22.860		2,860	475		3,335	3,825	
0200	48' x 9' high, with fixed panels	↓	.90	17.780		5,500	370		5,870	6,575	
0500	For bronze finish, add					15%					
350 0010	STAINLESS STEEL And glass entrance unit, narrow stiles										350
0020	3' x 7' opening, including hardware, minimum	2 Sswk	1.60	10	Opng.	1,430	220		1,650	1,925	
0050	Average		1.40	11.430		2,650	255		2,905	3,325	
0100	Maximum	↓	1.20	13.330		3,650	295		3,945	4,475	
1000	For solid bronze entrance units, statuary finish, add					60%					
1100	Without statuary finish, add				↓	45%					
400 0010	STOREFRONT SYSTEMS Aluminum frame, clear ⅜" plate glass,										
0020	incl. 3' x 7' door with hardware (400 sq. ft. max. wall)										
0500	Wall height to 12' high, commercial grade	2 Glaz	150	.107	S.F.	12.25	2.21		14.46	16.65	
0600	Institutional grade		130	.123		14.30	2.55		16.85	19.40	
0700	Monumental grade		115	.139		19.30	2.89		22.19	25	
1000	6' x 7' door with hardware, commercial grade		135	.119		15.40	2.46		17.86	20	
1100	Institutional grade		115	.139		18.55	2.89		21.44	25	
1200	Monumental grade	↓	100	.160		25.40	3.32		28.72	33	
1500	For bronze anodized finish, add					15%					
1600	For black anodized finish, add					20%					
1700	For stainless steel framing, add to monumental				↓	75%					
2000	For individual doors see Division 8.1										
600 0010	SWING DOORS Alum. entrance, 6' x 7', incl. hdwre & oper.	2 Sswk	.70	22.860	Opng.	4,770	505		5,275	6,050	600
0020	For anodized finish, add				"	400			400	440	

8.5	Metal Windows	CREW	DAILY OUTPUT	MAN-HOURS	UNIT	BARE COSTS				TOTAL	
						MAT.	LABOR	EQUIP.	TOTAL	INCL O&P	
100 0010	ALUMINUM SASH Stock, grade 2, glaze & trim not incl., casement	2 Sswk	200	.080	S.F.	10.75	1.77		12.52	14.65	100
0050	Double hung		200	.080		6.50	1.77		8.27	9.95	
0100	Fixed casement		200	.080		5	1.77		6.77	8.30	
0150	Picture window		200	.080		5.70	1.77		7.47	9.10	
0200	Projected window		200	.080		12.75	1.77		14.52	16.85	
0250	Single hung		200	.080		6	1.77		7.77	9.40	
0300	Sliding		200	.080	↓	8.20	1.77		9.97	11.85	
1000	Mullions for above, tubular		240	.067	L.F.	1.55	1.47		3.02	4.06	
2000	Custom aluminum sash, grade 3, glazing not included, minimum		200	.080	S.F.	15.70	1.77		17.47	20	
2100	Maximum	↓	85	.188	"	21	4.16		25.16	30	
200 0010	ALUMINUM WINDOWS Including frame and glazing, grade 2										200
1000	Stock units, casement, 3'-1" x 3'-2" opening	2 Sswk	10	1.600	Ea.	168	35		203	240	
1050	Add for storms					31			31	34	
1600	Projected, with screen, 3'-1" x 3'-2" opening	2 Sswk	10	1.600	↓	117	35		152	185	

85

Figure 8.9

Interior openings are defined as follows: framing material, glass, size and intermediate framing members. Components for each system include gasket setting or glazing bead and typical wood blocking.

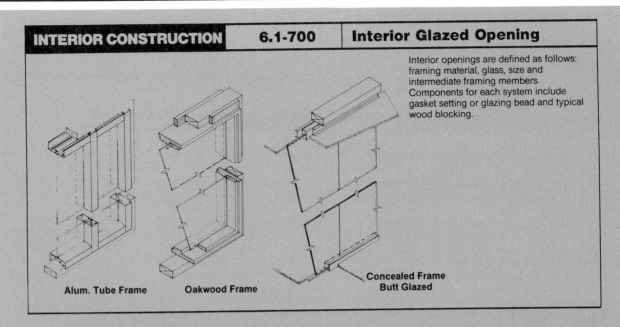

Alum. Tube Frame   Oakwood Frame   Concealed Frame Butt Glazed

System Components	QUANTITY	UNIT	COST PER OPNG.		
			MAT.	INST.	TOTAL
SYSTEM 06.1-700-1000					
GLAZED OPENING, ALUMINUM TUBE FRAME, FLUSH, ¼″ FLOAT GLASS, 6′ X 4′					
Aluminum tube frame, flush, anodized bronze, head & jamb	16.500	L.F.	76.23	98.67	174.90
Aluminum tube frame, flush, anodized bronze, open sill	7.000	L.F.	32.76	40.74	73.50
Joints for tube frame, clip type	4.000	Ea.	15.40		15.40
Gasket setting, add	20.000	L.F.	43		43
Wood blocking	8.000	B.F.	2.95	11.25	14.20
Float glass, ¼″ thick, clear, plain	24.000	S.F.	42.24	95.76	138
TOTAL			212.58	246.42	459

6.1-700			Interior Glazed Opening					
	FRAME	GLASS	OPENING-SIZE W X H	INTERMEDIATE MULLION	INTERMEDIATE HORIZONTAL	COST PER OPNG.		
						MAT.	INST.	TOTAL
1000	Aluminum flush	¼″ float	6′x4′	0	0	215	245	460
1040	Tube		12′x4′	3	0	515	530	1,045
1080			4′x5′	0	0	190	215	405
1120			8′x5′	1	0	355	395	750
1160			12′x5′	2	0	520	575	1,095
1240		⅜″ float	9′x6′	2	0	630	660	1,290
1280			4′x8′-6″	0	1	405	425	830
1320			16′x10′	3	1	1,675	1,750	3,425
1400		¼″ tempered	6′x4′	0	0	270	245	515
1440			12′x4′	3	0	630	525	1,155
1480			4′x5′	0	0	240	210	450
1520			8′x5′	1	0	450	395	845
1560			12′x5′	2	0	665	575	1,240
1640		⅜″ tempered	9′x6′	2	0	995	660	1,655
1680			4′x8′-6″	0	0	580	400	980
1720			16′x10′	3	0	2,550	1,625	4,175
1800		¼″ one way mirror	6′x4′	0	0	460	245	705
1840			12′x4′	3	0	1,000	520	1,520
1880			4′x5′	0	0	395	210	605
1920			8′x5′	1	0	770	390	1,160

296

Figure 8.10

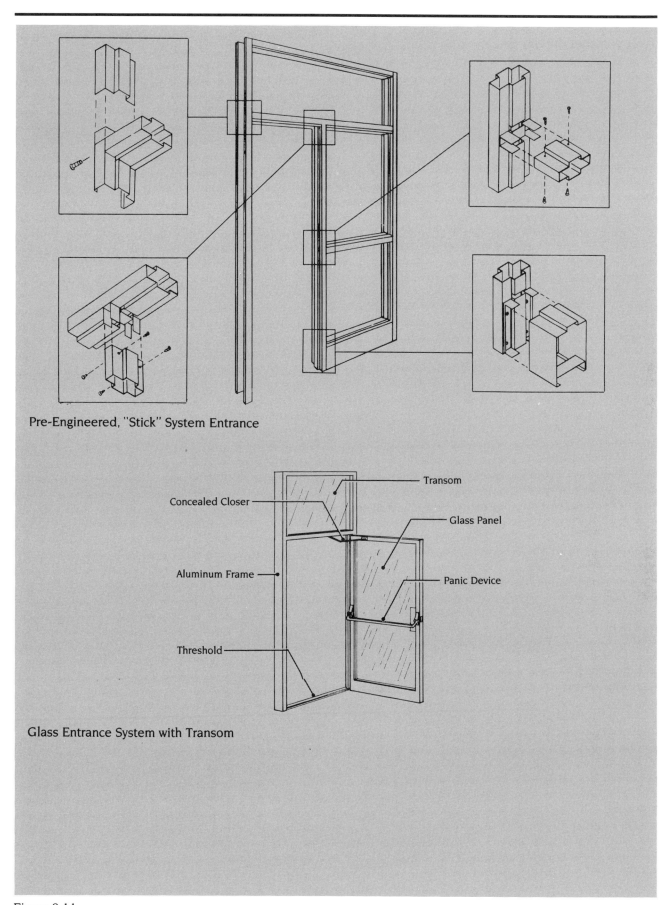

Pre-Engineered, "Stick" System Entrance

Glass Entrance System with Transom

Figure 8.11

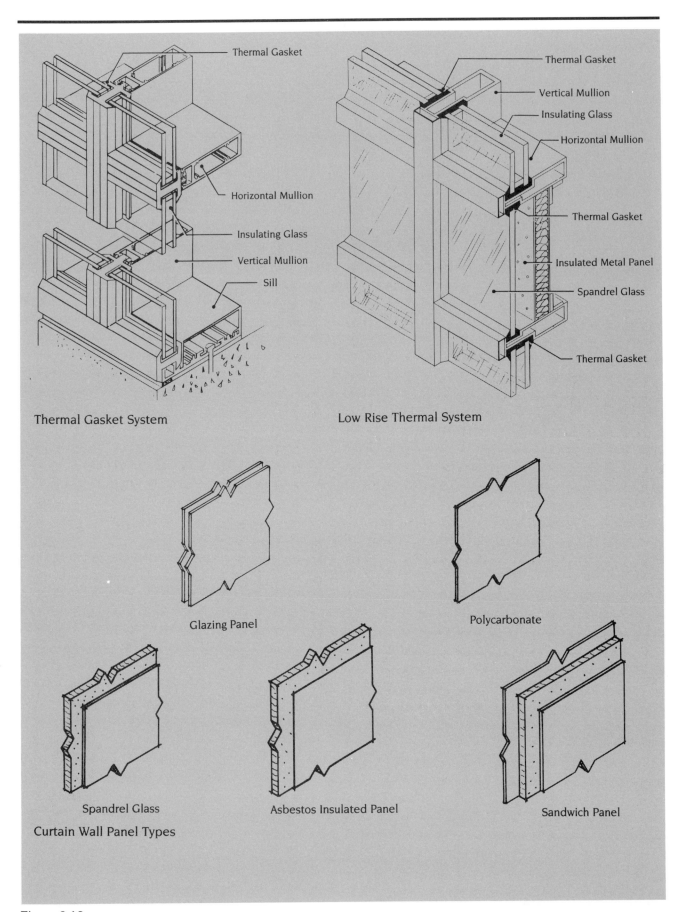

Thermal Gasket

Horizontal Mullion

Insulating Glass

Vertical Mullion

Sill

Thermal Gasket System

Thermal Gasket

Vertical Mullion

Insulating Glass

Horizontal Mullion

Thermal Gasket

Insulated Metal Panel

Spandrel Glass

Thermal Gasket

Low Rise Thermal System

Glazing Panel

Polycarbonate

Spandrel Glass

Asbestos Insulated Panel

Sandwich Panel

Curtain Wall Panel Types

Figure 8.12

Finish Hardware

Finish hardware is the construction industry term for the devices used to operate doors, windows, drawers, shutters, closets, and cabinets. This category includes such items as hinges, latches, locks, panic devices, security and detection systems, astragals, and weather stripping (some of which are shown in Figure 8.13a and b). In a typical building, the finish hardware accounts for between 2% and 3% of the total cost. Consequently, the difference between economy and quality hardware can mean a 1% difference in the total building cost.

A hardware specialist will often prepare a schedule and specify the hardware that is to be used for each opening. There are two general classfications of hardware: *Builder's*, and *Commercial*. Builder's is generally used for residential construction.

Another hardware specification is frequency of use, labelled heavy, light, or medium. The location where the hardware will be used is one more consideration. The size, weight, and material of a door and its frame, for example, will dictate the size and number of hinges. Building codes and security requirements will determine the selection of the proper fire barrier and electronic hardware in a modern building.

Most finish hardware is made of cast, wrought, or forged metal, such as iron, brass, bronze, white metal, or aluminum. These metals can be natural or plated and can be finished in dull, polished, Japaned, or antique finishes.

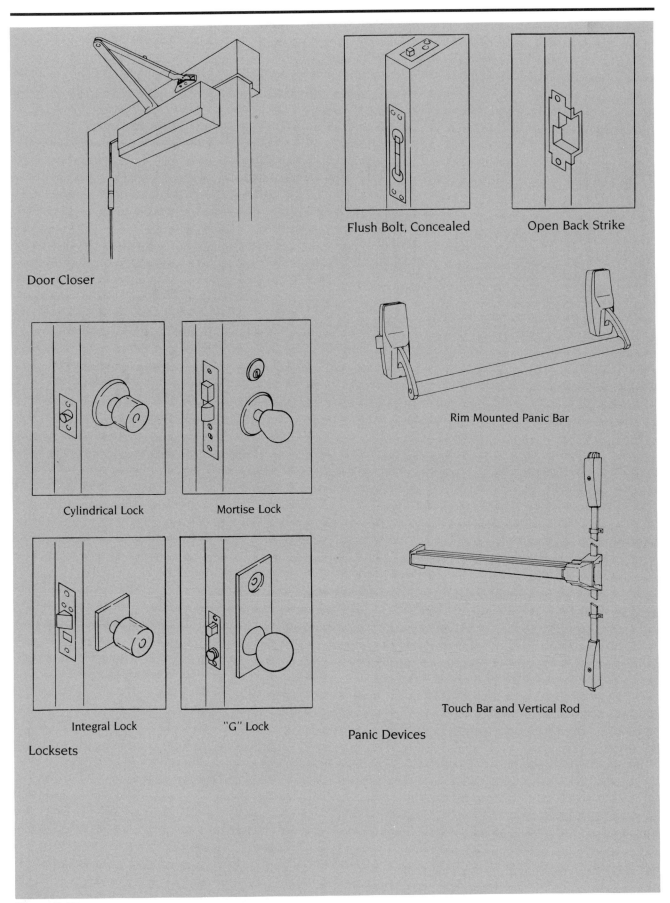

Door Closer

Flush Bolt, Concealed

Open Back Strike

Cylindrical Lock

Mortise Lock

Rim Mounted Panic Bar

Integral Lock

"G" Lock

Locksets

Panic Devices

Touch Bar and Vertical Rod

Figure 8.13a

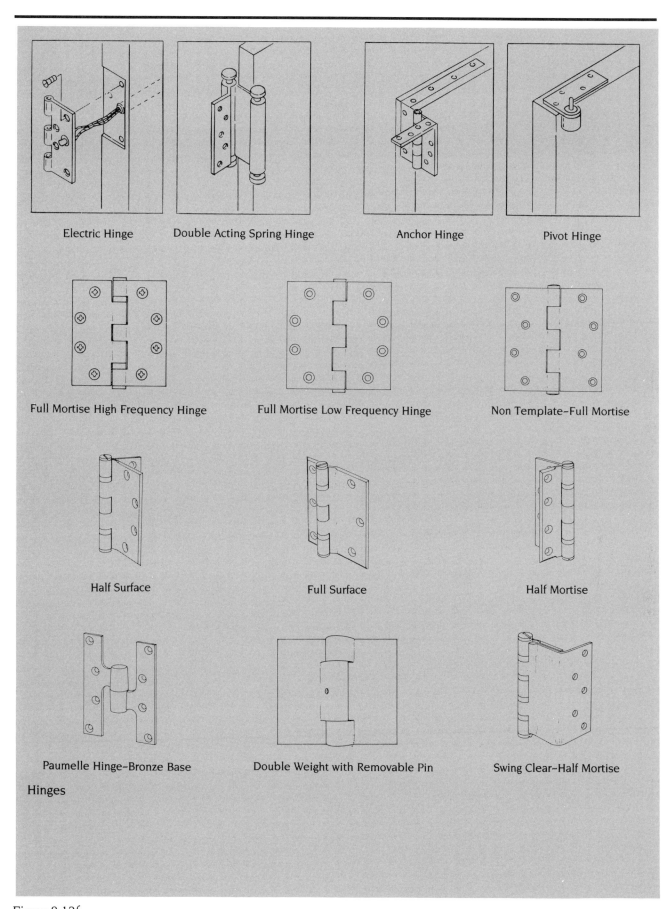

Electric Hinge   Double Acting Spring Hinge   Anchor Hinge   Pivot Hinge

Full Mortise High Frequency Hinge   Full Mortise Low Frequency Hinge   Non Template–Full Mortise

Half Surface   Full Surface   Half Mortise

Paumelle Hinge–Bronze Base   Double Weight with Removable Pin   Swing Clear–Half Mortise

Hinges

Figure 8.13b

Chapter 9
FINISHES

This chapter contains descriptions of the basic types of interior finishes with appropriate estimating methods. Materials used to finish walls, ceilings, doors, windows, and trimwork can include any of a number of products available. Most building codes (and specifications) require strict adherence to maximum fire, flame spread, and smoke generation characteristics. In today's fireproof and fire resistant types of construction, some finish materials may be the only combustibles used in a building project. Thus, these materials may have to be treated for fire retardancy, at an additional cost. The estimator must be sure that all materials meet the specified requirements and that any fire proofing necessary be included in the estimate for interior finishes.

Lath and Plaster

The different types of plaster work require varied pricing strategies. Large open areas of continuous walls or ceilings will require considerably less labor per unit of area than small areas or intricate work such as archways, curved walls, cornices, and window returns. Gypsum and metal lath are most often used as subbases, however, plaster may be applied directly on masonry, concrete, and, in some restoration work, wood. In the latter cases, a bonding agent may be specified. Illustrations of plaster partitions and metal lath are shown in Figure 9.1.

The number of coats of plaster may also vary. Traditionally, a scratch coat is applied to the substrate. A brown coat is then applied two days later, and the finish, smooth coat seven days after the brown coat. Currently, the systems most often used are two-coat and one-coat (imperial plaster on "blueboard"). Textured surfaces, with and without patterns, may be required. All of these variables in plaster work make it difficult to develop "system" prices. Each project, and even areas within each project, must be examined individually.

The quantity takeoff should proceed in the normal construction sequence — furring (or studs), lath, plaster and accessories. Studs, furring and/or ceiling suspension systems, whether wood or steel, should be taken off separately. Responsibility for the installation of these items should be made clear. Depending upon local work practices, lathers may or may not install studs or furring. These materials are usually estimated by the piece or linear foot, and sometimes by the square foot. Lath is traditionally estimated by the square yard for both gypsum and metal lath and is done more recently by the square foot. Usually, a 5% allowance for waste is included. Casing bead, corner bead, and other accessories are measured by the linear foot. An extra foot of surface area should be allowed for each linear foot of corner or stop. Although wood plaster grounds are usually installed by carpenters, they should be measured when taking off the plaster requirements.

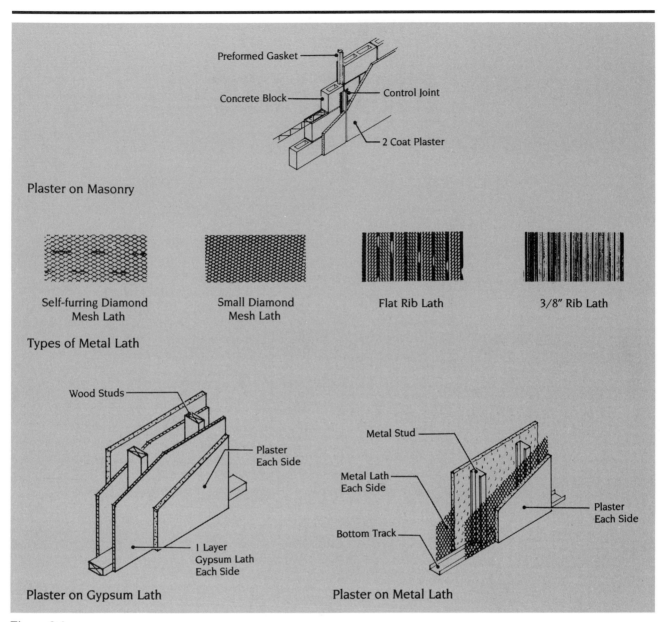

Preformed Gasket

Concrete Block

Control Joint

2 Coat Plaster

Plaster on Masonry

Self-furring Diamond
Mesh Lath

Small Diamond
Mesh Lath

Flat Rib Lath

3/8" Rib Lath

Types of Metal Lath

Wood Studs

Plaster
Each Side

1 Layer
Gypsum Lath
Each Side

Plaster on Gypsum Lath

Metal Stud

Metal Lath
Each Side

Bottom Track

Plaster
Each Side

Plaster on Metal Lath

Figure 9.1

Plastering is also traditionally measured by the square yard. Deductions for openings vary by preference – from zero deduction to 50% of all openings over 2 feet in width. Some estimators deduct a percentage of the total yardage for openings. The estimator should allow one extra square foot of wall area for each linear foot of inside or outside corner located below the ceiling level. The areas of small radius work should be doubled. Quantities are determined by measuring surface area (walls, ceilings, etc.). The estimator must consider both the complexity and the intricacy of the work, and in pricing plaster work, should also consider quality. Basically, there are two quality categories:

- *Ordinary* quality is used for commercial purposes, and with waves 1/8" to 3/16" in 10 feet. Angles and corner are fairly true.
- *First quality* with variations less than 1/16" in 10 feet. Labor costs for first quality work are approximately 20% more than that for ordinary plastering.

Drywall

With the advent of light gauge metal framing, tin snips are as important to the carpenter as the circular saw. Metal studs and framing are usually installed by the drywall subcontractor. The estimator should make sure that studs (and other framing – whether metal or wood) are not included twice by different subcontractors. In some drywall systems, such as shaftwall, the framing is integral and installed simultaneously with the drywall panels. Typical drywall partitions systems are shown in Figure 9.2.

Metal studs are manufactured in various widths (1-5/8", 2-1/2", 3-5/8", 4" and 6") and in various gauges, or metal thicknesses. They may be used for both load-bearing and non-load-bearing partitions, depending on design criteria and code requirements. Metal framing is particularly useful due to the prohibitive use of structural wood (combustible) materials in new building construction. Metal studs, tract and accessories are purchased by the linear foot, and usually stocked in 8' to 16' lengths, by 2' increments. For large orders, metal studs can be purchased in any length up to 20'.

For estimating, light gauge metal framing is taken off by the linear foot or by the square foot of wall area of each type. Different wall types – with different stud widths, stud spacing, or drywall requirements – should each be taken off separately, especially if estimating by the square foot.

Metal studs can be installed very quickly. Depending upon the specification, metal studs may have to be fastened to the track with self-tapping screws, tack welds, or clips, or may not have to be prefastened. Each condition will affect the labor costs. Fasteners such as self-tapping screws, clips, and powder-actuated studs are very expensive, though labor-saving. These costs must be included.

Drywall may be purchased in various thicknesses – 1/4" to 1" – and in various sizes – 2' x 8' to 4' x 20'. Different types include standard, fire resistant, water resistant, blueboard, coreboard, and pre-finished. There are many variables and possible combinations of sizes and types. While the installation cost of 5/8" standard drywall may be the same as that of 5/8" fire resistant drywall, the two types (and all other types) should be taken off separately. The takeoff will be used for purchasing, and material costs will vary.

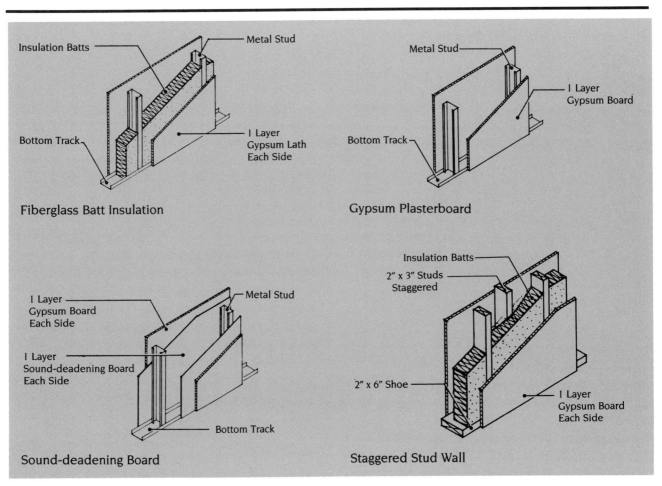

Fiberglass Batt Insulation

Insulation Batts — Metal Stud

Bottom Track — 1 Layer Gypsum Lath Each Side

Gypsum Plasterboard

Metal Stud — 1 Layer Gypsum Board

Bottom Track

Sound-deadening Board

1 Layer Gypsum Board Each Side — Metal Stud

1 Layer Sound-deadening Board Each Side

Bottom Track

Staggered Stud Wall

Insulation Batts

2" x 3" Studs Staggered

2" x 6" Shoe — 1 Layer Gypsum Board Each Side

Figure 9.2

Because drywall is used in such large quantities, current, local prices should always be checked. A variation of a few cents per square foot can become many thousands of dollars over a whole project.

Fire resistant drywall provides an excellent design advantage in creating relatively lightweight, easy to install firewalls (as opposed to masonry walls). As with any type of drywall partition, the variations are numerous. The estimator must be very careful to take off the appropriate firewalls exactly as specified. Even more important, the contractor must build the firewalls exactly as specified. Liabilities can be great. For example, a metal stud partition with two layers of 1/2" fire resistant drywall on each side may constitute a two-hour partition (when all other requirements such as staggered joints, taping, sealing openings, etc., are met). If a one-hour partition is called for, the estimator cannot assume that one layer of 1/2" fire resistant drywall on each side of a metal stud partition will suffice. Alone, it does not. When left to choose the appropriate assembly (given the rating required), the estimator must be sure that the system has been tested and approved for use – by Underwriters Laboratory as well as local building and fire codes and responsible authorities. In all cases, the drywall (and studs) for firewalls must extend completely from the deck below to the underside of the deck above, covering the area above and around any and all obstructions. All penetrations must be protected.

In the past, structural members – such as beams or columns – to be fireproofed had to be "wrapped" with a specified number of layers of fire resistant drywall – a very labor intensive and expensive task. With the advent of spray-on fireproofing, structural members can be much more easily protected. This type of work is usually performed by a specialty subcontractor. Takeoff and pricing are done by square foot of surface area.

When walls are specified for minimal sound transfer, the same continuous, unbroken construction is required. Sound proofing specifications may include additional accessories and related work. Resilient channels attached to studs, mineral fiber batts and staggered studs may all be used. In order to develop high noise reduction coefficients, double stud walls may be required with sheet lead between double or triple layers of drywall. (Sheet lead may also be required for X-ray installations.) Caulking is required at all joints and seams. All openings must be specially framed with double, "broken" door and window jambs.

Cavity shaftwall, developed for a distinct design advantage, is another drywall assembly which should be estimated separately. Firewalls require protection from both sides. Shaftwall, used at vertical openings (elevators, utility chases), can be installed completely from one side. Special track, studs (C-H or double E type) and drywall (usually 1" thick and 2' wide coreboard) are used and should be priced separately from other drywall partition components. Figure 9.3 illustrates several types of cavity shaftwall.

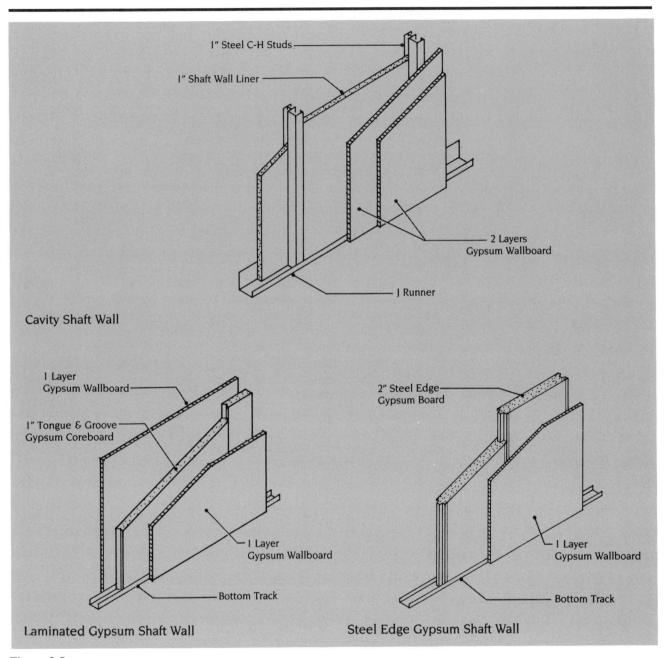

1″ Steel C-H Studs

1″ Shaft Wall Liner

2 Layers
Gypsum Wallboard

J Runner

Cavity Shaft Wall

1 Layer
Gypsum Wallboard

1″ Tongue & Groove
Gypsum Coreboard

1 Layer
Gypsum Wallboard

Bottom Track

Laminated Gypsum Shaft Wall

2″ Steel Edge
Gypsum Board

1 Layer
Gypsum Wallboard

Bottom Track

Steel Edge Gypsum Shaft Wall

Figure 9.3

Because of the size and weight of drywall, costs for material handling and loading should be included. Using larger sheets (manufactured up to 4' x 20') may require less taping and finishing, but these sheets may each weigh well in excess of 100 pounds and are awkward to handle. The weight of drywall must also be considered (and distributed) when loading a job on elevated slabs, so that allowable floor loads are not exceeded. All material handling involves costs that must be included.

As with plaster work, open spans of drywall should be priced differently from small, intricate areas that require much cutting and piecing. Similarly, areas with many corners or curves will require higher finishing costs than open walls or ceilings. Corners, both inside and outside, should be estimated by the linear foot, in addition to the square feet of surface area.

Although difficult because of variations, the estimator may be able to develop historical systems or assemblies prices for metal studs, drywall, taping and finishing, similar to those shown in Figure 9.4. When using systems, whether complete or partial, the estimator must be sure that the system, as specified, is exactly the same as the system for which the costs are developed. For example, a cost is developed for 5/8" fire resistant drywall, taped and finished. The project specifications require a firewall with two layers of the 5/8" drywall on each side of the studs. It could be easy to use the "system" cost for each layer, when only one of the two layers is to be taped and finished. The more detailed the breakdown and delineation, the less chance for error.

Tiles

Ceramic tiles and tiles manufactured from other materials, because of their modular installation and multi-purpose capabilities, provide an endless source of interior wall, floor, and counter top coverings. Standard tiles and the methods employed in setting them are designed and manufactured for the intended use of the covering. Many manufacturers can supply special shapes, complimentary items, and, especially bathroom accessories to complete the tile installation. Some of the more commonly applied uses of ceramic and other tiles include: toilet rooms, tubs, steam rooms, swimming pools, and other related installations where easily cleaned, water repellant, and durable surfaces are required.

Ceramic and other tiles are manufactured from several basic materials in a wide range of shapes, sizes, and finishes. Ceramic tiles are manufactured from clay, porcelain, or cement. Metal and plastic are the most commonly used materials for tiles not labelled as ceramic. In addition to standard ceramic tile materials, floor tiles may also include split brick pavers, quarry tiles, and terra cotta tiles. Tile shapes and sizes vary from 1" squares to variously sized rectangles, mosaics, patterned combinations, hexagons, octagons, valencia, wedges, and circles. They are available in multi-colored designs and mural sets or with individual pieces that are embossed with designs or pictures. Tiles with special designs and logos may also be custom manufactured. Standard and custom ceramic tiles are produced with glazed and unglazed surfaces with bright, matte, non-slip, and textured finishes.

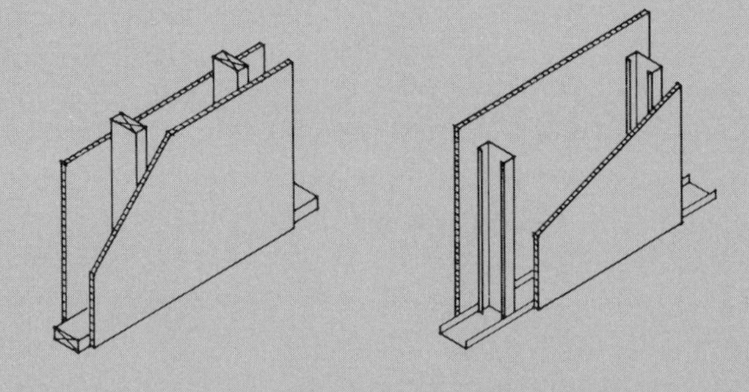

The drywall partitions/stud framing systems are defined by type of drywall and number of layers, type and spacing of stud framing, and treatment on the opposite face. Components include taping and finishing.

Cost differences between regular and fire resistant dry wall are negligible, and terminology is interchangeable. In some cases fiberglass insulation batts are included for additional sound deadening.

System Components	QUANTITY	UNIT	COST PER S.F.		
			MAT.	INST.	TOTAL
SYSTEM 06.1-510-1250					
DRYWALL PARTITION,⅝″ F.R.1 SIDE,⅝″ REG.1 SIDE,2″X4″STUDS16″ O.C.					
Gypsum plasterboard, nailed/screwed to studs, ⅝″F.R. fire resistant	1.000	S.F.	.32	.28	.60
Gypsum plasterboard, nailed/screwed to studs, ⅝″ regular	1.000	S.F.	.29	.28	.57
Taping and finishing joints	2.000	S.F.	.04	.48	.52
Framing, 2 x 4 studs @ 16″ O.C., 10′ high	1.000	S.F.	.32	.51	.83
TOTAL			.97	1.55	2.52

6.1-510				**Drywall Partitions/Wood Stud Framing**				
						COST PER S.F.		
	FACE LAYER	BASE LAYER	FRAMING	OPPOSITE FACE	INSULATION	MAT.	INST.	TOTAL
1200	⅝″ FR drywall	none	2 x 4, @ 16″ O.C.	same	0	1	1.55	2.55
1250				⅝″ reg. drywall	0	.97	1.55	2.52
1300				nothing	0	.66	1.03	1.69
1400		¼″ SD gypsum	2 x 4 @ 16″ O.C.	same	1-½″ fiberglass	1.75	2.31	4.06
1450				⅝″ FR drywall	1-½″ fiberglass	1.57	2.05	3.62
1500				nothing	1-½″ fiberglass	1.23	1.53	2.76
1600		resil. channels	2 x 4 @ 16″, O.C.	same	1-½″ fiberglass	1.59	2.91	4.50
1650				⅝″ FR drywall	1-½″ fiberglass	1.49	2.35	3.84
1700				nothing	1-½″ fiberglass	1.15	1.83	2.98
1800		⅝″ FR drywall	2 x 4 @ 24″ O.C.	same	0	1.57	2.01	3.58
1850				⅝″ FR drywall	0	1.25	1.73	2.98
1900				nothing	0	.91	1.21	2.12
2050			on 6″ plate	⅝″ FR drywall	0	1.59	2.09	3.68
2100				nothing	0	1.22	1.52	2.74
2200		⅝″ FR drywall	2 rows-2 x 4	same	2″ fiberglass	2.46	2.85	5.31
2250			16″O.C.	⅝″ FR drywall	2″ fiberglass	2.14	2.57	4.71
2300				nothing	2″ fiberglass	1.80	2.05	3.85
2400	⅝″ WR drywall	none	2 x 4, @ 16″ O.C.	same	0	1.10	1.57	2.67
2450				⅝″ FR drywall	0	1.05	1.56	2.61
2500				nothing	0	.71	1.04	1.75
2600		⅝″ FR drywall	2 x 4, @ 24″ O.C.	same	0	1.67	2.03	3.70
2650				⅝″ FR drywall	0	1.30	1.74	3.04
2700				nothing	0	.96	1.22	2.18
2800	⅝″ VF drywall	none	2 x 4, @ 16″ O.C.	same	0	1.80	1.43	3.23
2850				⅝″ FR drywall	0	1.40	1.49	2.89
2900				nothing	0	1.06	.97	2.03

291

Figure 9.4

Proper installation methods for ceramic tiles are determined by the location and the type of backing surface, which may vary from stud or masonry walls to wood or concrete floors. The tiling material may be placed in individual pieces, or it may be installed in factory-prepared back-mounted and ungrouted sections or sheets which cover two square feet or more as a unit. Tiling material is also available in factory prepared sections and patterns in which the tiles have been preset and pre-grouted with silicone rubber.

The two recognized methods of installation are the thick set, or mud set, method and the thin set method, in which the tile is directly adhered to the base, sub-base, or wall material. In *thick set floor installations*, Portland cement and sand are placed and screeded to a thickness of 3/4" to 1-1/4". For *thick set wall installations*, Portland cement, sand, and lime are placed on the backing surface and troweled to a thickness of 3/4" to 1'. With both floor and wall placement, the mortar may be reinforced with metal lath or mesh and can be backed with impervious membranes. The tiles may be placed on and adhered to the mortar bed while it remains plastic, or they may be placed after the bed has cured, adhered with a thin bond coat of Portland cement with sand additives. The thin set method of tile installation requires specially prepared mortars and adhesives on properly prepared surfaces. Latex Portland cement, epoxy mortar, epoxy emulsion mortar, epoxy adhesive, and organic adhesives are some of these specialized thin set preparations.

The grouting of the placed and adhered tile material is the final critical step in the installation process, as it ensures the sealing of the joints between the tiles and affects the appearance of the installation. The choice of grouting material to be used depends on the type of tile material and the conditions of its exposure. Some of the available grouting mixtures include Portland cement grouts with additives, mastic grout, furan resin grout, epoxy grout, and silicone rubber grout. The recommendations of the manufacturer should be carefully followed in all aspects of the grouting operation. A typical ceramic tile installation is shown in Figure 9.5.

Tile is usually recorded and priced per square foot. Linear features such as bullnose trim and cove base are estimated per linear foot. Specialties (accessories) are taken off as each.

Ceilings

There are three basic types of ceilings: exposed structural systems that are painted or stained, directly applied acoustical tile, gypsum board or plaster, or suspended systems, hung below the supporting superstructure to allow space for passage of ductwork, piping, and lighting systems.

Suspension systems usually consist of aluminum or steel main runners of light, intermediate or heavy duty classification, spaced 2, 3, 4 or 5 feet on center with snap in cross tees usually available in 1 to 5 foot lengths. Main runners are usually hung from the supporting structure with tie wire or metal straps spaced as required by the system. The runners and cross tees may be exposed, natural or painted, or concealed (enclosed by the ceiling tiles). Drywall suspension systems consist of main and cross tee members with flanges to allow the gypsum board to be screwed to the supports.

Acoustical ceilings are available in mineral fiber tiles with many patterns and textures. Sizes vary from 1 foot square to 2' x 5' rectangles. The face may be perforated, fissured, textured, or plastic covered. Ceiling tiles are also available in perforated steel, stainless steel, or aluminum to allow for easy cleaning in high humidity or kitchen areas. To retain the acoustical characteristics, these panels are filled with sound-absorbing pads. Specialty metal systems are also available as linear ceilings in varying widths in many finishes and configurations.

Integrated ceiling systems combine a suspension grid, air handling, lighting, and ceiling tiles into one modular system and have the benefit of a single source of responsibility.

Drywall ceilings may be applied directly to furring strips attached to the supporting structure or a compatible suspension system. They may be painted, sprayed with acoustical material, or covered with acoustical tiles. Plaster ceilings can be applied directly on self-furring or gypsum lath. Suspended plaster ceilings attach to cold rolled channels hung from the supporting structure and covered with metal or gypsum lath. Figures 9.6a and 9.6b show typical ceiling applications.

Fire resistance (in hours) for the various ceiling systems are evaluated as part of the floor or floor structural and ceiling assembly. To achieve a rating in hours, fire dampers, light fixtures, grilles, and diffusers are evaluated along with the structure, suspension system, and ceiling.

Sound barriers for acoustical deadening usually consist of materials with mass that deflects sound. They may be masonry, metal with sandwich insulation, leaded vinyl, fiberglass batts or sheet lead.

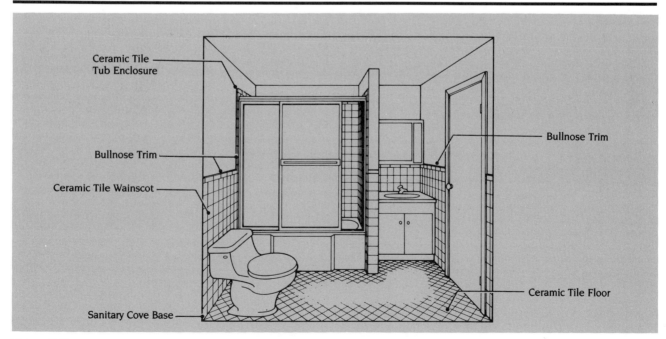

Figure 9.5

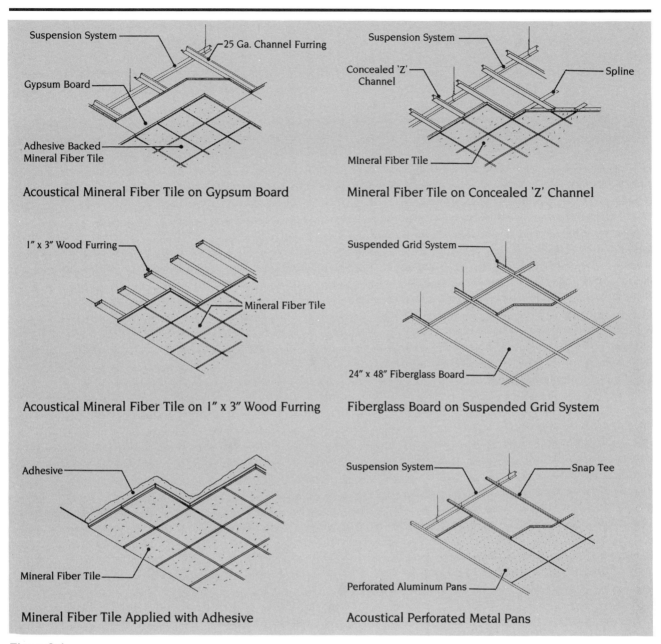

Suspension System — 25 Ga. Channel Furring

Gypsum Board

Adhesive Backed Mineral Fiber Tile

Acoustical Mineral Fiber Tile on Gypsum Board

1" x 3" Wood Furring

Mineral Fiber Tile

Acoustical Mineral Fiber Tile on 1" x 3" Wood Furring

Adhesive

Mineral Fiber Tile

Mineral Fiber Tile Applied with Adhesive

Suspension System — Spline

Concealed 'Z' Channel

Mineral Fiber Tile

Mineral Fiber Tile on Concealed 'Z' Channel

Suspended Grid System

24" x 48" Fiberglass Board

Fiberglass Board on Suspended Grid System

Suspension System — Snap Tee

Perforated Aluminum Pans

Acoustical Perforated Metal Pans

Figure 9.6a

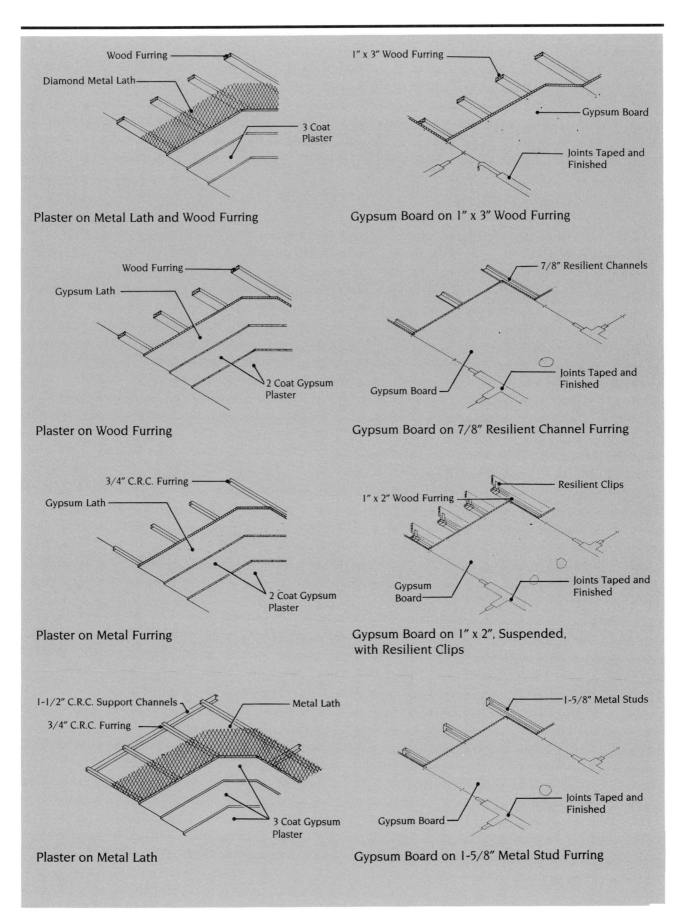

Plaster on Metal Lath and Wood Furring

Gypsum Board on 1" x 3" Wood Furring

Plaster on Wood Furring

Gypsum Board on 7/8" Resilient Channel Furring

Plaster on Metal Furring

Gypsum Board on 1" x 2", Suspended, with Resilient Clips

Plaster on Metal Lath

Gypsum Board on 1-5/8" Metal Stud Furring

Figure 9.6b

Suspended ceilings of acoustical tiles, drywall, or plaster are usually priced per S.F. Some system costs include all or part of the suspension grid, while others include only the panels and the suspension costs must be estimated separately. Patterned and angled ceilings require more material and labor to install.

There is little waste when installing grid systems (usually less than 5%), because pieces can be butted and joined. Waste for tile, however, can be as low as 5% for large open areas, to as high as 30%–40% for small areas and rooms. Waste for tile may depend on grid layout as well as room dimensions. Figure 9.7 demonstrates that for the same size room, the layout of a typical 2' x 4' grid has a significant effect on generated waste of ceiling tile. Since most textures and patterns on ceiling tile are aligned in one direction, pieces cannot be turned 90 degrees (to the specified alignment) to reduce waste.

Certain tile types, such as regular (recessed) require extra labor for cutting and fabrication at edge moldings. Soffits, fascias, and "boxouts" should be estimated separately due to extra labor, material waste, and special attachment techniques. Costs should also be added for unusually high numbers of tiles to be specially cut for items such as sprinkler heads, diffusers, or telepoles.

While the weight of ceiling tile is not the primary consideration that it is with drywall, some material handling and storage costs will still be incurred and must be included. Acoustical tile is very bulky and cumbersome, as well as fragile, and must be protected from damage before installation.

Carpeting

Carpet is often installed as a flooring material in office, commercial, and residential buildings because it provides an attractive, comfortable, and sound-softening pattern. Carpeting may be placed in large or small areas with relatively quick and clean installation.

Since carpeting is manufactured in various combinations of materials and textures, the plans should be read carefully; the type of carpet specified can affect the total price. For commercial carpeting, rolled goods are available in 54" and 12' widths, and the standard size of carpet tiles is 18" X 18" or 24" X 24". Carpet may be tufted, woven (axminister, wilton, or knitted), or fusion bonded. The surface texture may be level loop or multi-level loop, cut level pile, velvet cut pile, or a combination of cut and loop pile. Materials may be wool, nylon (with different brand name types), or acrylic. Carpets may be of one fabric or a blend.

If a custom color or pattern is required and the quantity exceeds 100 S.Y., there is generally no additional cost for manufacturing, provided the manufacturer has the appropriate equipment to construct the carpet. If the quantity of carpet required is less than 100 S.Y., and is a custom pattern or color, there may be an additional charge of up to 20%.

Carpet material may be specified or measured by its face weight in ounces per square yard, its pile height, its density, or stitches per inch. A commonly used formula for determining the carpet's density is:

$$\text{Density (oz./S.Y.)} = \frac{\text{Face weight, oz./S.Y. x 36}}{\text{Pile height, in.}}$$

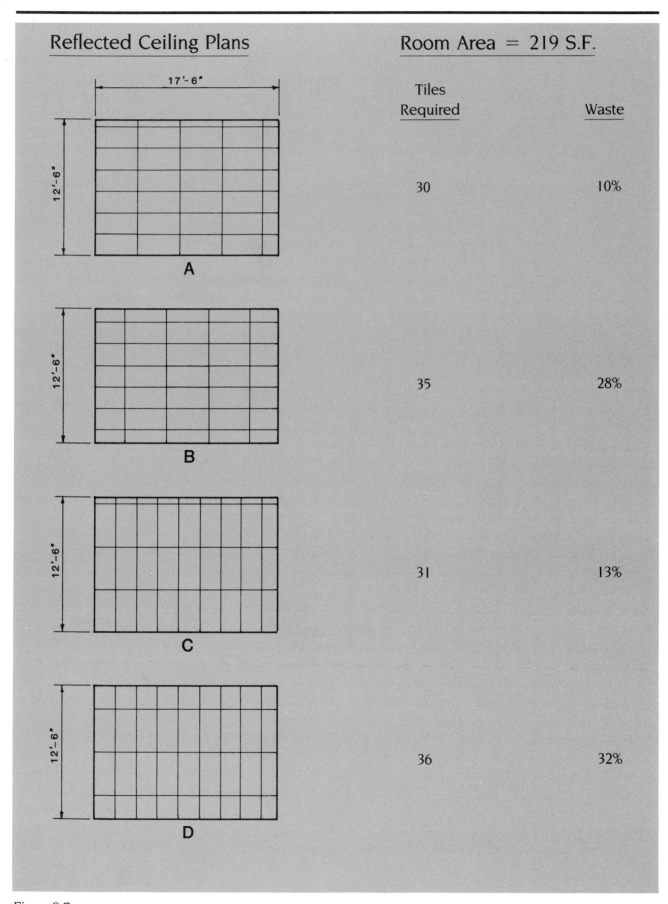

Reflected Ceiling Plans

Room Area = 219 S.F.

	Tiles Required	Waste
A	30	10%
B	35	28%
C	31	13%
D	36	32%

Figure 9.7

96

Tufted carpets are usually available in 12 or 15 foot widths. Woven carpets are manufactured in variable widths. A variety of backing materials, including polypropelene, jute, cotton, rayon, and polyester, provide support for the surface material. The methods used for integrating the surface materials and attaching them to the backing include tufting, weaving, fusion bonding, knitting, and needle punching. A second backing material of foam, urethane, or sponge rubber may be applied to the standard carpet backing to provide an attached pad.

The method of installation is usually determined by such variables as the size of the area to be covered, the amount and type of traffic it will carry, and the style of carpeting. Depending on these conditions, carpet may be installed over separate felt, sponge rubber, or urethane foam pads, or it may be directly adhered to the sub-floor. Wide expanses and carpet locations which carry heavy or wheeled traffic usually require direct glue-down installation to prevent wrinkling, bulging, and movement of the carpet. If padding is required in a glue-down installation, it must consist of an attached pad of synthetic latex, polyurethane foam, or similar material. Carpet may also be installed over a separate pad and then stretched and edge fastened. This method of installation is not recommended for large areas or heavy traffic situations.

When *rolled* or *broadloom carpet* with a large pattern repeat is specified, up to 15% additional carpet should be ordered to allow for pattern matching. The average waste for pattern matching is 10%–12%. To determine the quantity of carpet actually required, the estimator should prepare a seam diagram. This involves using an actual scale floor plan and showing where all the carpet seams will occur when the carpet is rolled on the floor. Waste will inevitably occur where there are cut outs or protruding corners, however, the diagram will provide an accurate determination of the amount of carpet that must be ordered, and acceptable locations for seams. To estimate the total square yards of carpet required, the following equation can be used, based on the total lineal feet of carpet measured on the seam diagram:

$$\frac{\text{Total L.F. of rolled carpet x 12}}{9} = \text{Total S.Y. of carpet required}$$

With the development of flat undercarpet cable systems for power, telephone, and data line wiring, the use of *modular carpet tiles* has increased greatly. Modular carpet tiles, when placed directly over the undercarpet cable run, provide easy access to the wiring system. Because taps can be made at any point along the route of the undercarpet system, facilities relocation can be accomplished conveniently, quickly, and economically.

Sizes for carpet tiles are exact. Since they fit together with factory edges, which are clean, the seam does not show. For this reason, there is less waste with carpet tiles than with broadloom or rolled carpet. However, 7 to 10 percent extra should be added to ensure proper color matching for future replacement.

To minimize waste due to dye lot variation, all carpeting should be ordered at the same time. The carpet should also be installed in the numerical sequence as identified on the rolls by the carpet manufacturer.

Composition Flooring

Composition floors are seamless coverings which are based on epoxy, polyester, acrylic, and polyurethane resins. Various aggregates or color chips may be combined with the resins to produce decorative effects, or special coatings may be applied to enhance the functional effectiveness of a floor's surface. Composition floors are normally installed in special-use situations where other flooring materials do not meet sanitary, safety, or durability standards. Special-use applications of composition flooring include: manufacturing areas, food processing areas, kitchens, beverage bottling plants, parking facilities, multi-purpose rooms, gymnasiums, locker rooms, and explosion hazard areas.

In most cases, the materials which are combined to form the composition floor are determined by and designed for the specific application. Some of the available composition flooring systems include: terrazzo (for decorative floors), wood toppings and sealers, concrete toppings and sealers, waterproof and chemical resistant floors, gymnasium sports surfaces, interior and exterior sports surfaces, and conductive floors. The materials employed in these systems are applied by trowel, roller, squeegee, notched trowel, spray gun, brush, or similar tools. After the flooring materials have been applied and allowed to cure, they may be sanded and/or sealed to produce the desired texture and surface finish. Composition floors are normally installed by trained and approved specialty contractors.

Resilient Floors

Resilient floors are designed for situations where durability and low maintenance are primary considerations. The various resilient flooring materials and their installation formats include: asphalt tiles, cork tiles, polyethylene in rolls, polyvinyl chloride in sheets, rubber in tiles and sheets, vinyl composition tiles, and vinyl in sheets and tiles. All of these materials may be manufactured with or without resilient backing and, except for the polyethylene rolls, they are available in a wide range of colors, designs, textures, compositions, and styles. Rubber and vinyl accessories, which are designed to compliment any type of flooring material, include: bases, beveled edges, thresholds, corner guards, stair treads, risers, nosings, and stringer covers. Examples are illustrated in Figure 9.8.

The manufacturer's recommendations should be carefully followed for the detailed aspects of the installation of any resilient floor. Generally, any concrete floor surface should be dry, clean, and free from depressions and adhered droppings. Curing and separating compounds should also be thoroughly removed from the surface before the resilient floor is placed. Special consideration is required for installations on slabs or wood surfaces that are located below grade level or above low crawl spaces.

The floor thickness specified is important to note, as there are different thickness grades for commercial and residential use. Resilient floors are usually estimated and priced per square foot; base and stair stringers are priced per linear foot.

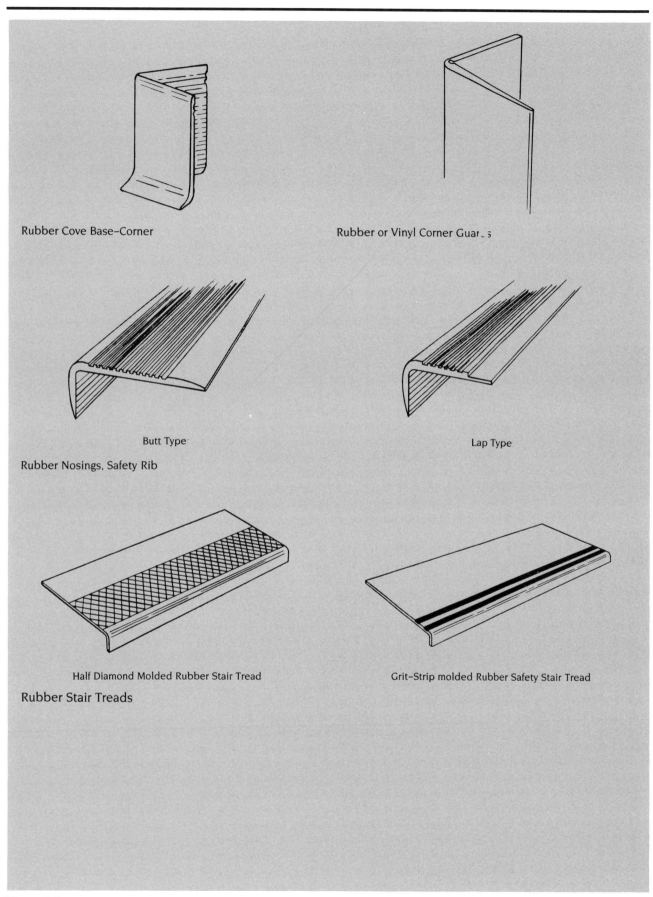

Rubber Cove Base–Corner

Rubber or Vinyl Corner Guards

Butt Type

Lap Type

Rubber Nosings, Safety Rib

Half Diamond Molded Rubber Stair Tread

Grit–Strip molded Rubber Safety Stair Tread

Rubber Stair Treads

Figure 9.8

Due to the thinness and flexibility of resilient goods, defects in the subfloor easily "telegraph" through the material. Consequently, the subfloor material and the quality of surface preparation are very important. Subcontractors will often make contracts conditional on a smooth, level subfloor. Surface preparation, which can involve chipping and patching, grinding or washing, is often an "extra". Some surface preparation will invariably be required and an allowance for this work should be included in the estimate. This is especially true in renovation where, in extreme cases, the floor may have to be leveled with a complete application of special lightweight leveling concrete.

Terrazzo

Terrazzo flooring materials provide many options to produce colorful, durable, and easily cleaned floors. Conventional ground and polished terrazzo floors employ granite, marble, glass, and onyx chips in a choice of specialized matrices. Well-graded gravel and other stone materials may be used to create different textures when added to the mix. Precast terrazzo tiles and bases, which are normally installed finished and polished on a cement sand base, are also available in many color combinations and aggregates.

Conventional installation of terrazzo flooring involves the mixing of the aggregate with one of three commonly used bonding matrices and the placing of the mix in sections defined by divider strips. The matrices employed to bond the aggregate include: cementitious matrices, which consist of natural or colored Portland cement; cement with an acrylic additive; and resinous matrices, which consist of epoxy and polyester. The divider strips, which are manufactured from zinc, brass, or colored plastic, provide for expansion and are, therefore, positioned over breaks in the substrate and at critical locations where movement is expected in large sections of flooring. The divider strips are also used to terminate pours, to act as leveling guides, and to permit changes in the aggregate mix, section-to-section, to create designs or patterns.

The terrazzo flooring materials may be installed on a cement sand underbed, or may be applied in the thin set method directly to the concrete slab. Sand-cushioned terrazzo employs three layers of material: a 1/4" thick sand cushion placed on the slab and covered by an isolation membrane; a mesh-reinforced underbed of approximately 1-3/4" in thickness; and a 1/2" thick terrazzo topping (see Figure 9.9). Bonded terrazzo consists of two layers of material: a 1-3/4" thick underbed placed on the slab and a 1/2" thick terrazzo topping. Monolithic terrazzo flooring is comprised of a single 1/2" thick layer of terrazzo topping applied directly to the slab after control joints have been saw cut into the slab and the divider strips grouted into place. With the thin set method, the terrazzo mix is applied in a single 1/4" thick layer directly on the concrete slab. The surface of stair treads and risers may be poured in place with a 1/2" thick layer of terrazzo topping on a 3/4" underbed, or they may be installed as precast terrazzo pieces on a 3/4" thick underbed.

Wood Floors

Wood flooring material is manufactured in three common formats: solid or laminated planks; solid or laminated parquet; and end-grain blocks, which are available in individual pieces or in pre-assembled strips. Commonly employed hardwood flooring materials include ash, beech, cherry, mahogany, oak, pecan, teak, and walnut, as well as exotic hardwood species, such as ebony, karpa wood, rosewood, and zebra wood. Cedar, fir, pine, and spruce are among the popular softwood flooring materials. End-grain block flooring is manufactured from alder, fir, hemlock, mesquite, oak, and yellow pine.

Strip flooring material is supplied in several different milling formats and combinations, including tongue-and-grooved and matched, square-edged, and jointed with square edges and splines. The installation of tongue-and-groove flooring usually requires blind nailing or fastening by means of metal attachment clips, while square-edged flooring is usually face fastened. Strip floors may be installed over wood-framed subfloors of planks or plywood sheets which have been fastened to the floor joists. A layer of building paper should be laid between the subfloor and the finish strip floor to provide a comfort cushion and to reduce noise. If the strip floor is placed over a concrete surface, a subfloor of exterior plywood should be fastened directly to the slab and protected from moisture by polyethylene vapor barrier between the concrete and the subfloor. The plywood subfloor may also be fastened to lapped sleepers which have been imbedded in asphalt floor mastic on the slab and draped with the polyethylene vapor barrier before the subfloor is placed.

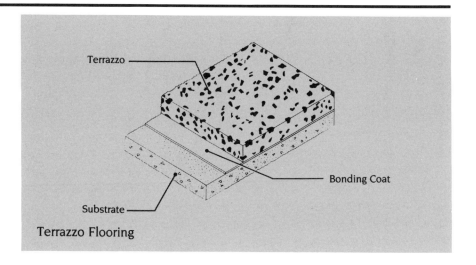

Terrazzo

Bonding Coat

Substrate

Terrazzo Flooring

Figure 9.9

Parquet floors are prefabricated in panels of various sizes which are milled with square edges, tongue-and-groove edges, or splines. The panels may also contain optional adhered backings which protect the flooring material from moisture, add comfort to the walking area, provide insulation, and deaden sound. Adhesives are normally employed to attach the parquet flooring panels to firm, level subfloors of concrete, wood, plywood, particleboard, resilient tile, or terrazzo. Some parquet floor manufacturers also supply feature strips and factory-fashioned moldings to cover the required expansion space at the edges of the floor.

Wood block floors for industrial application are manufactured in individual blocks and pre-assembled strips. The surface dimensions of individual blocks are 3" by 6", 4" by 6", and 4" by 8", with nominal thicknesses ranging from 1-1/2" to 4". The blocks are normally installed in a layer of pitch applied to a concrete floor. The finish coating of pitch or similar material is then squeezed into the joints between the blocks to provide additional fastening strength. A sealer is applied to provide surface finish. End-grain strip block flooring is placed in mastic adhesive which has been troweled into a dampproofing membrane that covers the concrete subfloor. After the strip block flooring material has been laid, it is sanded, filled, and finished with penetrating oil.

Wood gymnasium or sports floors usually require specialized installation methods because of their unique function and large size. They may be placed over sleepers that are installed on cushions or pads, or they may be laid directly over a resilient base material or plywood sub-base on cushioned pads. If the flooring material is placed on a sleeper support system, metal clips may be applied for fastening; if it is placed on a plywood subfloor, direct nailing is normally used for fastening. Because large floor areas, such as gymnasiums, require a wide expansion space at the edges, the placement and type of closure strip installed to cover the space deserve special consideration. Wood floors are usually priced per square foot of coverage required. Some typical wood floor installations are shown in Figure 9.10.

Painting and Finishing

Painting and finishing are required to protect interior and exterior surfaces against wear and corrosion and to provide a coordinated finish appearance on the protected material. Paints can be normally classified by their binders or vehicles. Alkyds are oil-modified resin used to manufacture fast-drying enamels. Chlorinated rubber produces coatings that are resistant to alkalies, acids, chemicals, and water. Catalyzed epoxies are two-part coatings that produce a hard film resistant to abrasion, traffic, chemicals, and cleaning. Epoxy esters are epoxies (modified with oil) that dry by oxidation, but are not as hard as catalyzed epoxies. Latex binders (commonly polyvinylacetate, acrylics, or vinyl acetate-acrylics) are binders mixed with a water base. Silicone alkyd binders are oil modified to produce coatings with heat resistance and high-gloss retention. Urethanes are isocynate polymers that are modified with drying oils or alkyds. Vinyl coating solutions are plasticized copolymers of vinyl chloride and vinyl acetate dissolved in strong solvents. Zinc coatings are primers high in zinc dust content dispersed in various vehicles to provide coatings to protect steel from oxidation.

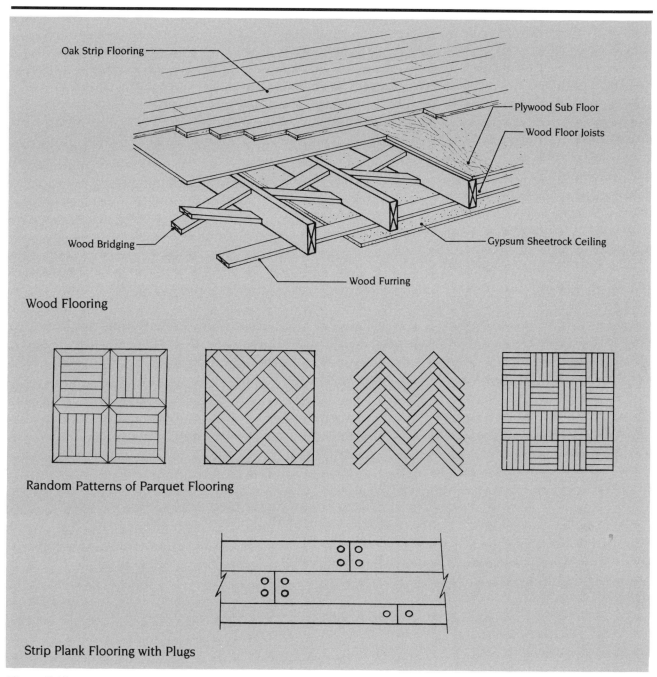

Oak Strip Flooring

Plywood Sub Floor

Wood Floor Joists

Wood Bridging

Gypsum Sheetrock Ceiling

Wood Furring

Wood Flooring

Random Patterns of Parquet Flooring

Strip Plank Flooring with Plugs

Figure 9.10

In order to achieve successful painting, the surface to be painted must first be cleaned. Preparation methods for painting include solvent cleaning, hand-tool cleaning, (including wire brushing, scraping, chipping, and sanding), power-tool cleaning, white metal blasting (including removal of all rust and scale), commercial blasting (to remove oil, grease, dirt, rust), and scaling or brushing off blast (to remove oil, grease, dirt, and loose rust). These surface preparation processes can be costly and must be included in the estimate where required.

Fillers, primers, and sealers are manufactured for use with paint or clear coatings for metal, wood, plaster, drywall, and masonry. Sanding fillers are available for clear coatings and block fillers for masonry construction.

Topcoats or finish coats should be specified and applied to meet service requirements. Finishes can be supplied for normal use and atmospheric conditions, hard service areas, and critical areas. Finishes can be furnished in gloss, semi-gloss, low lustre, eggshell, or flat. Intumescent coatings (expandable paints) are available to meet fire-retardant requirements.

As with most finishes, architects and designers will be particularly insistent about adherence to specifications and specified colors for painting. This is not an area in which to cut corners in estimating or performance of the work. The specifications usually clearly define acceptable materials, manufacturers, preparation, and application methods for each different type of surface to be painted. Many samples of various colors and/or finishes may be required for approval by the architect or owner before final decisions are made.

When estimating painting, the materials and methods are included in the specifications. Areas to be painted are usually defined on a Room Finish Schedule and taken off from the plans and elevations in square feet of surface area. Odd shaped and special items can be converted to an equivalent wall area. The table in Figure 9.11 includes suggested conversion factors for various types of surfaces.

While the factors in Figure 9.11 are used to determine quantities of equivalent wall surface areas, the appropriate, specified application method must be used for pricing.

The choice of application method will have a significant effect on the final cost. Spraying is very fast, but the costs of masking the areas to be protected may offset the savings. Oversized rollers may be used to increase production. Brushwork, on the other hand, is labor intensive. The specifications often include (or restrict) certain application methods. Typical coverage and man-hour rates are shown in Figure 9.12 for different application methods. These figures include masking and protection of adjacent surfaces.

Depending on local work rules, some unions require that painters be paid higher rates for spraying, and even for roller work. Higher rates also tend to apply for structural steel painting, for high work and for the application of fire retardant paints. The estimator should determine which restrictions may be encountered.

Balustrades:		1 Side x 4
Blinds:	Plain	Actual area x 2
	Slotted	Actual area x 4
Cabinets:	Including interior	Front area x 5
Downspouts and Gutters:		Actual area x 2
Drop Siding:		Actual area x 1.1
Cornices:	1 Story	Actual area x 2
	2 Story	Actual area x 3
	1 Story Ornamental	Actual area x 4
	2 Story Ornamental	Actual area x 6
Doors:	Flush	Actual area x 1.5
	Two panel	Actual area x 1.75
	Four Panel	Actual area x 2.0
	Six Panel	Actual area x 2.25
Door Trim:		LF x 0.5
Fences:	Chain Link	1 side x 3 for both sides
	Picket	1 side x 4 for both sides
Gratings:		1 side x 0.66
Grilles:	Plain	1 side x 2.0
	Lattice	Actual area x 2.0
Mouldings:	Under 12" Wide	1 SF/LF
Open Trusses:		Length x Depth x 2.5
Pipes:	Up to 4"	1 SF per LF
	4" to 8"	2 SF per LF
	8" to 12"	3 SF per LF
	12" to 16"	4 SF per LF
	Hangers Extra	
Radiators:		Face area x 7
Sanding and Puttying:	Quality Work	Actual area x 2
	Average Work	Actual area x 0.5%
	Industrial	Actual area x 0.25%
Shingle Siding:		Actual area x 1.5
Stairs:		No. of risers x 8 widths
Tie Rods:		2 SF per LF
Wainscoting, Paneled:		Actual area x 2
Walls and Ceilings:		Length x Width no deductions for less than 100 SF
Window Sash:		1 LF of part = 1 SF

Figure 9.11

Wall Coverings

Wall coverings are manufactured, printed, or woven in burlaps, jutes, weaves, grasses, paper, leather, vinyl, silks, stipples, corks, foils, sheets, cork tiles, flexible wood veneers, and flexible mirrors. Wallpaper, vinyl wall coverings, and woven coverings are usually available in different weights, backings, and quality. Surface preparation and adhesive selection are important considerations in placing wall coverings.

Wall coverings are usually estimated by the number of rolls. Single rolls contain approximately 36 S.F.; this figure forms the basis for determining the number of rolls required. Wall coverings are, however, usually sold in double or triple roll bolts.

The area to be covered is measured by its length times the height of the wall (above baseboards) in order to get the square footage of each wall. This figure is divided by 30 to obtain the number of single rolls, allowing 6 S.F. of waste per roll. One roll should be deducted for every two door openings. Two pounds of dry paste make about three gallons of ready-to-use adhesive and hangs about 36 single rolls of light to medium weight paper, or 14 rolls of heavyweight paper. Application labor costs vary with the quality, pattern, and type of joint required.

For vinyls and grass cloths requiring no pattern match, a waste allowance of 10% should be expected (approximately 3.5 S.F. per roll). Wall coverings which require a pattern match may have about 25%–30% waste, or 9–11 S.F. per roll. Waste can run as high as 50%–60% on wall coverings with a large, bold, or intricate pattern repeat.

Painting

Item	Coat	One Gallon Covers			In 8 Hrs. Man Covers			Man Hours per 100 S.F.		
		Brush	Roller	Spray	Brush	Roller	Spray	Brush	Roller	Spray
Paint wood siding	prime	275 S.F.	250 S.F.	325 S.F.	1150 S.F.	1400 S.F.	4000 S.F.	.695	.571	.200
	others	300	275	325	1600	2200	4000	.500	.364	.200
Paint exterior trim	prime	450	—	—	650	—	—	1.230	—	—
	1st	525	—	—	700	—	—	1.143	—	—
	2nd	575	—	—	750	—	—	1.067	—	—
Paint shingle siding	prime	300	285	335	1050	1700	2800	.763	.470	.286
	others	400	375	425	1200	2000	3200	.667	.400	.250
Stain shingle siding	1st	200	190	220	1200	1400	3200	.667	.571	.250
	2nd	300	275	325	1300	1700	4000	.615	.471	.200
Paint brick masonry	prime	200	150	175	850	1700	4000	.941	.471	.200
	1st	300	250	320	1200	2200	4400	.364	.364	.182
	2nd	375	340	400	1300	2400	4400	.615	.333	.182
Paint interior plaster or drywall	prime	450	425	550	1600	2500	4000	.500	.320	.200
	others	500	475	550	1400	3000	4000	.571	.267	.200
Paint interior doors and windows	prime	450	—	—	1300	—	—	.333	—	—
	1st	475	—	—	1150	—	—	.696	—	—
	2nd	500	—	—	1000	—	—	.800	—	—

Figure 9.12

Commercial wall coverings are available in widths from 21" to 54", and in lengths from 5-1/3 yards (single roll) to 100 yard bolts. To determine quantities, independent of width, the linear (perimeter) footage of walls to be covered should be measured. The linear footage should then be divided by the width of the goods, to determine the number of "strips" or drops. The number of strips per bolt or package can be determined by dividing the length per bolt by the ceiling height.

$$\frac{\text{Linear Footage of Walls}}{\text{Width of Goods}} = \text{No. of Strips Required}$$

$$\frac{\text{Length of Bolt (roll)}}{\text{Ceiling Height}} = \text{No. of Strips (whole no.) per Bolt (roll)}$$

Finally, divide the quantity of strips required by the number of strips per bolt (roll) in order to determine the required amount of material, using the same waste allowances as above.

$$\frac{\text{No. of Strips Required}}{\text{No. of Strips per Bolt (roll)}} = \text{No. of Bolts (rolls)}$$

Surface preparation costs for wall covering must also be included. If the wall covering is to be installed over new surfaces, the walls must be treated with a wall sizing, shellac, or primer coat for proper adhesion. For existing surfaces, scraping, patching and sanding may be necessary. Requirements will be included in the specifications.

Chapter 10
SPECIALTIES

Specialties include prefinished, manufactured items that are usually installed at the end of a project when other finish work is complete. Following is a partial list of items that may be included in Division 10 — Specialties, of the specifications.

- Bathroom accessories
- Bulletin Boards
- Chutes
- Control boards
- Directory boards
- Display cases
- Key Cabinets
- Lockers
- Mail boxes
- Medicine cabinets
- Partitions:
    - folding accordion
    - folding leaf
    - hospital
    - moveable office
    - operable
    - portable
    - shower
    - toilet
    - woven wire
- Part bins
- Projection screens
- Security gates
- Shelving
- Signs
- Telephone enclosures
- Turnstiles

A thorough review of the drawings and specifications is necessary to be sure that all items are accounted for. The estimator should list each type of item and the recommended manufacturers. Often, no substitutes are allowed. Each type of item is then counted. Takeoff units will vary with different items.

Quotations and bids should be solicited from local suppliers and specialty subcontractors. The estimator must include all appropriate shipping and handling costs. When a specialty item is particularly large, job-site equipment may be needed for placement or installation.

The estimator should pay particular attention to the construction requirements which are necessary to Division 10 work but are included in other divisions. Almost all items in Division 10 require some form of base or backing for proper installation. These requirements may or may not be included in the construction documents, but are usually listed in the manufacturers' recommendations for installation. It is often stated in the General Conditions of the specifications that "the contractor shall install all products according to manufacturers' recommendations" – a catch-all phrase that places responsibility on the contractor (and estimator).

Examples of such items are concrete bases for lockers and supports for ceiling-hung toilet partitions. If concrete bases for lockers cannot be installed, unit partitions are erected, or at least accurately laid out. The specific locker must be approved by the architect before the exact size of the base is determined. Installation of the concrete often requires a small truckload (hence an extra charge for a minimum order) and hand placement with wheelbarrow and shovel. Similarly, supports for toilet partitions cannot be installed until precise locations are determined. Such supports can be small steel beams, or large angles that must be welded in place.

Preparation costs prior to the installation of specialty items may, in some cases, exceed the costs of the items themselves. The estimator must visualize the installation in order to anticipate all of the work and costs.

Partitions

Folding partitions, operable walls, and relocatable partitions are manufactured in a variety of sizes, shapes, and finishes. Operating partitions include folding accordion, folding leaf, (both shown in Figure 10.1) or individual panel systems. These units may be operated by hand or power. Relocatable partitions include the portable type, which is designed for frequent relocation, and the demountable type, which is designed for infrequent relocation.

Operating partitions are supported by aluminum, steel, or wood framing members. The panels are usually filled with sound-insulation material, as most partitions are rated by their sound reduction qualities. Panel skin materials include aluminum, composition board, fabric, or wood. The panels may be painted or covered with carpeting, fabric, plastic laminate, vinyl, or wood paneling. Chalkboards and tackboards may be mounted on the skin of the paneling to add another dimension to their use. Panels may also be custom faced to designer's specifications. Large operating partitions are generally installed by factory specialists after the supporting members and framing have been supplied and erected by the building contractor.

Depending on the particular use, the amount of available space, and the design of the area to be partitioned, many different stacking methods and track configurations can be employed. Operating partitions may adopt a recessed-stacked, center-stacked, parallel-stacked, or exposed-stacked format. Track systems are available in straight lengths or curves, or with right angle layouts and switches. Operating partitions are usually top-track supported, but some models are available with floor supports.

Relocatable and portable panels are manufactured of the same materials used in operable walls. Vinyl-covered gypsum board may also be used as a face material. Partitions may be installed to ceiling height with adjustable top or bottom seals, or to varied heights with end-hinged supports. Manufacturer's details vary widely with various panel systems.

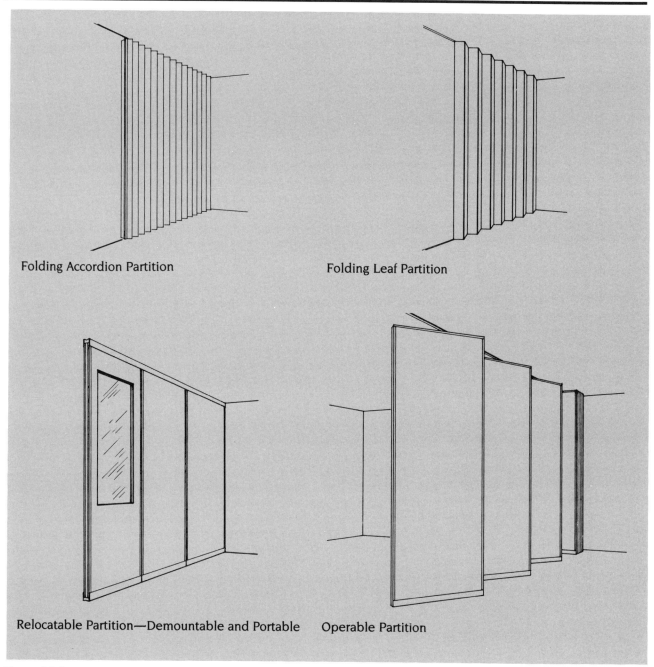

Folding Accordion Partition

Folding Leaf Partition

Relocatable Partition—Demountable and Portable

Operable Partition

Figure 10.1

Toilet Partitions, Dressing Compartments, and Screens

Toilet partitions, dressing compartments, and privacy screens are manufactured in a variety of materials, finishes, and colors. They are available in many stock sizes for both regular and handicapped-equipped water closets. These partitions may be custom fabricated to fit special size or use requirements, and may be supported from the floor, braced overhead, or hung from the ceiling or wall. Available finish materials include marble, painted metal, plastic laminate, porcelain enamel, and stainless steel. Various types of toilet partitions are shown in Figure 10.2.

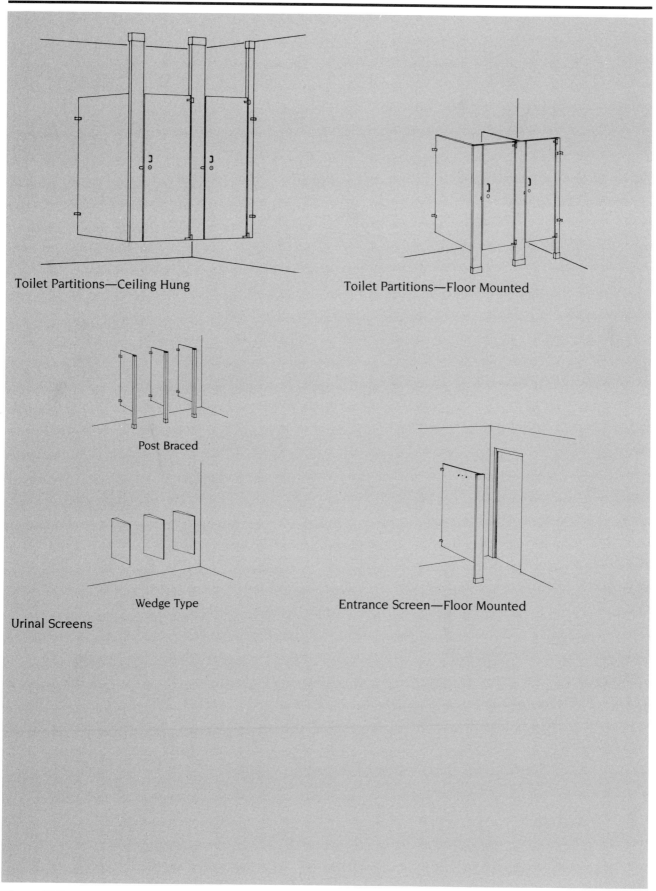

Toilet Partitions—Ceiling Hung

Toilet Partitions—Floor Mounted

Post Braced

Wedge Type

Urinal Screens

Entrance Screen—Floor Mounted

Figure 10.2

Chapter 11

ARCHITECTURAL EQUIPMENT

Architectural equipment includes the permanent fixtures that cause the space to function as designed — book stacks for libraries, vaults for banks. The construction documents may specify that the owner will purchase architectural equipment directly, and that the contractor will install it. In such cases, the estimator must include the installation cost of the equipment. If architectural equipment is furnished by the owner, it is common practice to add about 10% of the materials cost into the estimate. This procedure covers handling costs associated with the materials. Often the contractor is responsible for receipt, storage, and protection of these owner-purchased items until they are installed and accepted.

The following items might be included in this category:
- Appliances
- Bank equipment
- Baking equipment
- Church equipment
- Checkout counter
- Darkroom equipment
- Dental equipment
- Detention equipment
- Health club equipment
- Kitchen equipment
- Laboratory equipment
- Laundry equipment
- Medical equipment
- Movie equipment
- Refrigerator food cases
- Safe
- Sauna
- School equipment
- Stage equipment
- Steam bath
- Vocational shop equipment
- Waste handling equipment
- Wine vault

As with specialties, described in Chapter 10, architectural equipment must be evaluated to determine what is required from other divisions for its successful installation. Some possibilities are:

- Concrete work
- Miscellaneous metal supports
- Rough carpentry and backing
- Mechanical coordination
- Electrical requirements

Division 11 – Architectural Equipment includes equipment that can be packaged and delivered complete or partially assembled by the factory. Some items can or must be purchased and installed by an authorized factory representative. The estimator must investigate these variables in order to include adequate costs.

Chapter 12
FURNISHINGS

Numerous furnishings may be specified in the plans for commercial interiors projects. Some of these furnishings may be mass produced items, while others may be custom-made, one-of-a-kind items. Furnishings may be supplied by the owner, supplied by the owner and installed by the contractor, or carried under a specified allowance. Any of the following projects may require furnishings, and will be addressed in this chapter:

- Office Furniture
- Hotel Furniture
- Dormitory Furniture
- Hospital Furniture
- Library Furniture
- Restaurant Furniture
- School Furniture

In addition, components of interior furnishing systems, listed below, will be described:

- Stack and Folding Chairs
- Booths
- Multiple Seating
- Lecture Hall Seating
- Upholstery
- Folding Tables
- Files and File Systems
- Cabinets
- Blinds
- Shades
- Drapes
- Bedspreads
- Panels and Dividers
- Coat Racks and Wardrobes
- Mattresses and Box Springs
- Floor Mats
- Posts
- Lamps
- Plants
- Artwork
- Desk Accessories
- Ash/Trash Receivers

Before beginning a furniture estimate, the estimator should carefully review both the furnishings drawings and the complete set of construction documents and specifications. Items such as casework, wall art, or banquette seating in a restaurant may not be specified in the furnishings drawings, but will be found in the construction documents. A systematic method should be used to insure that all furnishing items are accounted for. The full set of drawings could be scanned and any item that may affect the furnishings costs marked with a colored pencil. Use of a standardized form such as a Means *Cost Analysis* form can also aid in accounting for all items in the project (shown in Figure 12.1).

A system for labeling each type or style component is also helpful. For example, all tables of a particular type or style could be labeled T-1, another style T-2, and yet another T-3. The estimator could then take off each style table specified, and record them by category and quantity. T-1:43 would mean there are 43 tables of the specific type and style denoted by T-1. This coding system is helpful both for quick reference and for accounting purposes.

Furniture can also be estimated using the Systems Estimating Method. This method involves the takeoff not by individual items but by workstations. These workstations may include raceways for electrical and communications requirements. Acoustical requirements and sizes of components vary with each manufacturer, as well as types of accessories required, such as task lights, bulletin boards, and coat racks.

To determine the costs, workstations should be priced by type, such as secretarial, management, or executive. Within each category there may be various configurations to meet specific user requirements. Costs similar to those in Means *Interior Cost Data* can be used for estimating. An example of the cost development for a workstation is shown in Figure 12.2. Note that a table and chairs which would most likely be included, must be estimated separately, as described above. To determine actual cost, each component in a system should be priced individually. This will determine the total cost per workstation. An additional percentage should be added to the total systems cost for installation.

Office Furniture

Office materials are generally made of wood, metal, or plastic laminate. Construction of drawer glides, detailing, and the configuration of drawers and compartments distinguishes the various products avilable. The estimator should verify whether what may look like identical desks from two manufacturers is really the same desk. The construction can differ greatly and may appreciably affect the cost.

Chairs vary in style and quality. Office seating is usually ergonomic in design, but each style may offer different features. The specifications should be carefully reviewed, as each style chair within a given price range may have unique options. For example, lounge chairs vary in price based on construction quality and covering fabric.

Tables for office use may range from specific name designer tables to generic drum and sled base tables. The design and style called for will determine competitiveness of bidding. Designer furniture, for example, items which are patented by a specific designer, may be available through one specific dealership. Generic items, on the other hand, such as sled base and drum tables, can be obtained through many furniture dealerships and bidding for these items may be competitive.

Means Forms

COST ANALYSIS

SHEET NO.

PROJECT _____ ESTIMATE NO. _____

ARCHITECT _____ DATE _____

TAKE OFF BY: _____ QUANTITIES BY: _____ PRICES BY: _____ EXTENSIONS BY: _____ CHECKED BY: _____

DESCRIPTION	SOURCE/DIMENSIONS			QUANTITY	UNIT	MATERIAL		LABOR	
						UNIT COST	TOTAL	UNIT COST	TOTAL

II.71

Figure 12.1

119

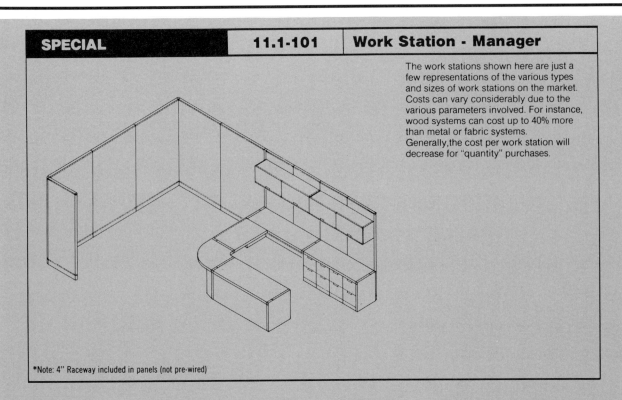

SPECIAL	11.1-101	Work Station - Manager

The work stations shown here are just a few representations of the various types and sizes of work stations on the market. Costs can vary considerably due to the various parameters involved. For instance, wood systems can cost up to 40% more than metal or fabric systems. Generally, the cost per work station will decrease for "quantity" purchases.

*Note: 4" Raceway included in panels (not pre-wired)

System Components	QUANTITY	UNIT	COST EACH		
			MAT.	INST.	TOTAL
SYSTEM 11.1-101					
WORK STATION - MANAGER					
Acoustical panel, aluminum frame, fabric covered, 48″ wide x 80″ high	5.000	Ea.	2130		2130
30″ wide x 80″ high	5.000	Ea.	1595		1595
End rail, aluminum	2.000	Ea.	149		149
Corner post, aluminum	2.000	Ea.	280		280
Junction rail, aluminum	1.000	Ea.	150		150
Panel connector, aluminum	6.000	Ea.	120		120
Overhead storage unit, 2 doors, 60″ wide, incl. brackets	2.000	Ea.	640		640
Lock for storage unit	2.000	Ea.	30		30
Overhead task light, 48″ long, w/30 watt fluorescent lamp, incl. bracket	2.000	Ea.	240		240
Hung pedestal with two 12″ deep drawers	4.000	Ea.	1320		1320
Lock for pedestal	4.000	Ea.	49		49
Worksurface, 20″ x 58-½″	1.000	Ea.	121		121
20″ x 60″, with short modesty panel	1.000	Ea.	171		171
24″ x 60″, with full modesty panel	1.000	Ea.	205		205
End panel support, 20″ wide	2.000	Ea.	273		273
24″ wide	1.000	Ea.	137		137
Outside leg support, 20″ wide	1.000	Ea.	137		137
Return worksurface, 24″ x 34-¼″, with full modesty panel	1.000	Ea.	133		133
Transition surface, 24″ radius, with full modesty panel	1.000	Ea.	358		358
Bracket for worksurface return	1.000	Ea.	29		29
Inner leg support, 24″ wide	2.000	Ea.	273		273
Installation cost	1.000	Ea.		150	150
TOTAL			8540.	150	8690

Figure 12.2

Conference tables are usually fabricated by specialty manufacturers. They may be made of wood, plastic laminate, or stone (marble or granite). Prices should be obtained from conference table manufacturers. There may be added costs for shipping, crating, and delivery, due to special size or special handling requirements. Bases for conference tables may vary greatly in type and style. Before adding the cost of shipping, delivery, and installation, the base specifications should be noted, since size can affect these costs significantly.

The quality, style, and price of office furnishings available may vary greatly, therefore, the estimator should determine exactly what is being specified by technical description if not by specific manufacturer. Any grey areas should be clarified by the specifier, as bidding can be competitive.

Hotel Furniture

Hotel furnishings include bed frames, headboards, mattresses, benches, desks, chairs, dressers, guest tables, mirrors, sleep sofas, cocktail and end tables. Specifications should be reviewed to determine requirements for each of these items. Large quantities of hotel furniture may be ordered from a guest room manufacturer, with quantity discounts if no special tooling is required. However, availability of the specified items should be verified, so that bids are not based on discontinued items. Substitute items could incur additional costs.

Restaurant Furniture

Restaurant furnishings can often be a competitive area. Since the cost of chairs and tables of all types, styles, sizes and finishes can vary greatly, the amount of detail specifed should first be determined. The quality of upholstered seating required and finishes specified may affect costs, especially if the fabric or finish is available from only a single manufacturer. Fabric may be COM (Customer's Own Material), ordered from a fabric manufacturer and delivered to a furniture manufacturer who covers a specific style chair with the material.

The cost of tables will vary considerably, depending on quality specified. Restaurant table tops and bases are generally sold separately, and therefore should be priced separately. The type of top and base required should be determined from the specifications. Top materials may be wood, glass, or marble, with various edge and top finishes. Bases may be made of wood, metal, or stone. Tops and bases are usually shipped separately, the top having predrilled holes for installation of the base. An additional cost for assembling the table to the base should also be considered when figuring installation cost.

Dormitory Furniture

Dormitory furniture includes all bookcases, beds, chairs, chests, desks, dressing units, mattresses, mirrors, night stands and wardrobes. These units are usually made of wood, plastic laminate, and/or metal, or combinations thereof.

When estimating the cost of dormitory furniture, material and installation requirements should be estimated separately. Additional material, such as hardware, or additional labor may be necessary to install these units. Beds may be stackable or bunkable; built-in lighting units may be required on hutch units; or special sized units may be called for.

Library Furniture

Library furniture includes book stacks, attendent desks, book display items, book trucks, card catalogues, study carrels both single and double face, chairs, charge desks, dictionary stands, exhibit cases, globe stands, newspaper and magazine racks, tables for card catalogue reference and study use. Bids for these types of furnishings should be obtained from a speciality contractor since there is special detailing required in many of the pieces of casework to meet the functional requirements for libraries.

School Furniture

School furniture includes chairs and desks made of molded plastic or wood and metal. Estimates should be obtained from local furniture dealers specializing in school furniture.

Furnishing Items

The following section describes items often encountered in most furniture estimates. These furnishings may be ordered through a commercial furniture dealership, an interior designer, or an architectural office with purchasing services. Since costs may vary greatly, quotations for specific items and quantities should be obtained. These quotes are often only good for a limited period of time, and any time limit should be noted.

Stack and Folding Chairs

Chairs may be metal, wood, or plastic. Before estimating the amount of chairs required, the amount that can be stored/stacked should be obtained from the manufacturer, since this will determine the number of dollys and carriers required. If upholstered chairs are specified, the fabric should be determined because this cost can be significant, especially for stacking chairs used for hotel ballroom projects since a large quantity is usually required. Floor guides for carpet and hard floors are different and this could affect the cost of the chairs. Whether tablet arms are required should be verified and if they are to be operable or fixed determined prior to estimating the cost. Chair finishes vary from lacquered wood to painted metal, therefore, the estimator should check the specifications for all finishes prior to determing cost.

Booths

Booths, also known as fixed seating, come in a variety of sizes and shapes. The cost will vary according to the type of seating specified. Banquette upholstered seating for restaurant use, for example, will differ in price from that of fixed seating, which consists only of one piece plastic chair and plastic laminated table top units such as those in fast food restaurants. When banquette seating is specified, the estimator should check the specifications and drawings to verify exactly what is being called for. In many cases the banquette seating should be estimated by a specialty contractor where curved units are required since these are not standard items.

Multiple Seating

Multiple seating is used in airports, hotel lobbies, and reception areas. Costs will vary depending on the construction specifications. Numerous types, styles, and materials are available, therefore, the specifications and drawings should be carefully reviewed to determine what is required.

Lecture Hall Seating

Lecture hall seating requires pedestal mounted (floor mounted) seats. The cost for this kind of seating will be affected by the following factors: the amount and quality of upholstery required; whether a veneer surface is called for; whether the tablet arm is to be retractable or fixed; and the type of material the shell is to be made of. The shell may be made of wood, metal, or plastic. Design and style are other considerations which may affect the cost of lecture hall seating.

Upholstery

Upholstery may be used in interior projects on chairs or as wallcovering. Upholstery materials include various blends of nylons, polyester, silk, or wool. To estimate the cost of upholstery, the specifications should be carefully reviewed to determine the type of weave, weight, color, pattern requirements and fabric width. There may be dye lot variations in various rolls, therefore, the quantity of material needed must be carefully estimated, to ensure that an adequate amount of material from the same dye lot is available for the job.

Fabric treatments may include flame retardant, acrylic backing, paper backing, and fabric treatment protection. When estimating the cost for fabric treatments, costs for testing of fabric/treatment compatibility should also be considered in the estimate.

Folding Tables

Folding tables are used in conference centers, hotels, schools, and wherever large assemblies of people are expected for dining or conference purposes. Shapes may be rectangular or circular. Sizes vary from conventional depths of 18", 30", and 36", and lengths of 48", 60", 72" and 96". Rounds are usually 60" and 72" diameters. Carts for storage and moving these items should be included in the estimate.

Files & File Systems

There are grades of filing systems for different uses, ranging from light residential to heavy administrative use. Files may be two to four drawer, vertical or lateral. The specifications should be carefully checked since the specifier has determined what type of files will be appropriate for each function. Floor loading capacities must be considered.

Where large, mobile filing systems are used with floor mounted tracks in filing banks, or for file systems on rollers or rotating banks, a specialty supplier and/or contractor should be consulted for accurate pricing. The price of these systems may vary and the speciality contractor will be experienced in installing file systems (and familiar with the costs). A structural engineer should be consulted whenever excessive loads are to be placed on a floor not specifically designed for such a purpose.

Cabinets

Cabinets for kitchen, hospital, and residential use are generally prefabricated in fixed sizes, ready for installation. Plastic laminated counter tops come in a variety of widths and may be stock or custom made. Prices for cabinets — wood, plastic laminate, and metal — can be taken from Means *Interior Cost Data*. Laminated tops are usually done by a speciality contractor skilled in assembling laminated products, since special post forming machines are required to make plastic laminated curved surfaces. Costs should be adjusted, however, when small quantities are ordered.

Blinds

Blinds may be either vertical or horizontal, in a variety of finishes including vinyl, metal, fabric, and wood. Slat sizes and type of operation may vary. The estimator should verify all options in the specifications. The finish for both sides should be checked, since the two sides may differ and this may affect the cost. Unique installations, such as door hung blinds which require "hold down clips" or blinds in sloped skylights which require special side rails to keep the blinds in place should be verified with a local speciality contractor to insure accurate estimates.

Shades

Shades include basswood rollup, insulative, solar screening fiberglass, and interior insulative shutters. To estimate the cost of shades, the square feet of shade required should first be determined, and this number should be multiplied by the material costs provided in Means *Interior Cost Data*. For irregular shaped windows (e.g., round or triangular shapes) estimates should be obtained from local speciality contractors.

Drapes

Drapes are manufactured in a variety of colors, finishes, and sizes. Since the windows in commercial projects are usually commercial sizes and not residential sizes, drapes are bid by specialty contractors. Drapes may be lined, or unlined, cotton or black out for light control, of varying fabrics. The weight of the fabric is also important, because of the significant effect on cost. Weighting may be required at the bottom of the drapes. The stitch pattern and finish required at the top and edges should also be verified. Generally a speciality contractor should be consulted to price exactly what is specified. In large projects, (e.g., hotels and performing arts centers) the specifer should be consulted to insure that all items are included in the project at the time of the estimate. Special flame retardent treatment is usually required. When flame treatment is called for, lab fire tests may have to be performed, and costs added, if necessary.

The method of installation may affect the cost of the drapes and will be indicated in the specifications. The method of installation required will vary, depending on whether the drapes are to be manual or automatic (motorized), or whether the drapes will be hung with a snap or hook attachment to the rod system. The specific operation of the drapes is also important. For example, one panel may be required for one-way draw, or two panels for central draw operation. Manual drapes can be operated from either a pull wand or cord.

Bedspreads

Bedspreads for hospitality projects are generally fitted or throw style. They may be filled or unfilled with a lining, either cotton or synthetic. The type of fabric required generally determines the cost of the bedspread, depending on the style, color, and pattern. The speciality contractor accomplishing the drapes usually supplies the bedspreads as well, therefore pricing can be confirmed for bedspreads by the same contractor.

Panels and Dividers

Panels and dividers for office use vary in type, style, and size. They may be powered or non-powered, with a fabric or hard edge of metal or wood. Acoustical rating will vary with each manufacturer, as well as available colors and types of connectors.

Coat Racks & Wardrobes

Coat racks and wardrobes are usually wood or metal, with interior components which can vary from project to project. The specifications should be checked to determine the height, width, number of hangers required, whether they are to be included in the price or are extra, and the amount of hat storage required above. Hat storage is indicated by the number and width of shelves at the top of the unit. Finishes should be verified. If custom colors are required, a speciality manufacturer should be consulted as there may be a surcharge for custom colors.

Mattresses and Box Springs
If there are special construction requirements, the cost for mattresses and box springs should be obtained directly from a manufacturer for the accurate estimates. Commercial grades differ from residential, therefore, the specification requirements should be given to the manufacturer.

Floor Mats
Floor mats are usually made to order. Size, material, and installation should be considered when estimating. Material for floor mats may be recessed units made of rubber, aluminum, or synthetics. Floor mat sizes should be confirmed and field-measured prior to ordering. Tile units, for example, come in boxes with a certain amount per box. The specific manufacturer should be consulted to determine the number of boxes required. Installation procedures may vary, from mechanical fasteners to glue, and should be confirmed with the manufacturer.

Posts
Posts are generally used for crowd control in theaters and banks. The costs are determined based on the type of post (simple or decorative) and the type and legnth of rope required.

Lamps
Lamps may be ceramic, glass, metal, and/or stone, ceiling, wall or floor mounted. The specifications will indicate the finishes, colors, and types of lampshades required, as this can affect the cost of the items. Lamps are shipped without light bulbs, and these should be ordered separately. Installation of wall mounted lamps in hotel projects may require special clips or bolts. In this case, an additional cost for special bolts should be included in the estimate. A special color may be specified to match the project design which necessitates that the lamps be obtained from one specific manufacturer. Quantity discounts may be available for large quantities.

Plants
Permanent plants for commercial use indicated in Means *Interior Cost Data* are permanent fabric plants. An additional cost for fill (e.g., bark or rock) should be included in the estimate for plants. An additional 10% should generally be added for shipping and crating charges for permanent plants.

Planters and containers may be made of fiberglass, metal, or wood. There may be added costs for packing and crating. Liners, if specified, will also add to the plant cost.

Art Work
Commercial art work is generally one of three types: photography, poster, or reproduction. Glass or plastic may be called for as a cover material. The cost will be significantly higher for art work with glass as the cover material since there may be additional packing charges. To reduce the cost, the pictures could be shipped to a local framer who would insert the glass locally. These arrangements, however, have to be agreed upon with the contractor and client prior to estimating. For one-of-a kind or unique pieces, quotes must be obtained from a specific broker or art gallery.

Desk Accessories

Desk accessories include ash trays, book ends, calendars with pads, carafes, desk pads, pen sets, letter trays and memo boxes. There are numerous possible finishes and colors for these items, including polished metal, stone (e.g., marble) and wood finishes. The estimator should check the specifications and drawings to determine what is required. Specific desk accessories are generally furnished by only one specific manufacturer.

Ash/Trash Receivers

Trash receivers come in a variety of shapes and sizes with various options. Ash/Trash receivers may be simple sand urns or units with mechanical self cleaning tops. The estimator should check the furnishings specifications for the type of operation required and options, as well as finishes, as they will have an impact on the cost of the unit.

Packing, Crating, Insurance, Shipping, and Delivery

There is often an added cost for packing, crating, insurance, shipping and delivery of furnishings. This added cost should be anticipated for each furnishings item where applicable. The specifications will dictate how the furniture is to be packed and crated. Otherwise large pieces of furniture might be blanket wrapped at no additional cost and put on a truck without the extra protection of a crate or proper packing. Large pieces such as hotel furniture, and office items like chairs and desks, should be packed and crated. The extra charge varies depending on the size of the piece of furniture, since crating and additional insurance charges are extra and are carried on items which are of "high value". The specifications should be checked for this requirement.

Shipping costs for furnishings can generally be estimated at 10% to 15% of the furniture cost. Furnishings may be delivered FOB (Freight on Board) Designation and are drop shipped at the receiving dock.

Delivery time can vary from project to project. Delivery time for small, stock orders will be 30 to 120 days. For medium size office and hotel projects, or for custom orders, delivery can be from 60 to 120 days. Therefore, for each furniture item, delivery time should be anticipated. Many manufacturers dealing in office furnishings offer a quick ship/delivery program, but this is usually only for certain basic stock items, offered in limited quantities since they do not usually store large inventories of office furniture.

Delivery charges are generally extra. For example, if furnishings are for the fourth floor of an office building, the contract may only be for delivery to the loading dock of the address specified. Provisions for storage and handling should be made ahead of time for large quantities of furnishings and the costs appropriately included.

Chapter 13
SPECIAL
CONSTRUCTION

A partial list of items in the special construction section, Division 13, of the specifications and drawings appears below. Subcontractor quotations should be carefully evaluated to make sure all the required items are included. Some of the materials may have to be supplied and/or installation performed by other suppliers or subcontractors.

- Acoustical enclosures
- Air curtains
- Integrated ceilings
- Pedestal access floors
- Refrigerators – walk-in
- Shielding
- Sports courts
- Vault front

It is a good idea to review this portion of the project with the subcontractor to determine both the exact scope of the work and those items that are not covered by the quotation. If the subcontractor requires services such as excavation, unloading, or other temporary work, then these otherwise excluded items must be included elsewhere in the estimate.

Often, manufacturers of these products require that only trained personnel install the product. Otherwise, material and performance warrantees may be voided.

The specialty subcontractor will have more detailed information at hand concerning the system. The more detailed the estimator's knowledge of a system, the easier it may be subdivided into material, labor, and equipment costs that will fully identify the direct cost of the specialty item for future purposes.

Chapter 14
CONVEYING
SYSTEMS

Conveying systems applied to interiors may include, but are not limited to, the following:

- Correspondence lifts
- Dumbwaiters
- Elevators
- Escalators
- Material handling conveyers
- Motorized car distribution systems
- Moving ramps and walks
- Parcel lifts
- Pneumatic tube systems
- Vertical conveyers

Current quotations from a competent contractor should be obtained for all of the items listed above, if specified and shown on the plans. Budget costs are available and should, when used properly, allow sufficient money to cover material and installation costs for new buildings. Installation in existing buildings must be priced on each individual project. Hydraulic elevators with telescoping shafts have now been marketed to combat some of the problems encountered when installing elevators in existing buildings. A checklist for the various systems is shown below.

Dumbwaiters

- capacity
- floors
- speed
- size
- stops
- finish

Elevators

- hydraulic or electric
- capacity
- floors
- stops
- finish
- door type
- special requirements
- geared or gearless
- size
- number required
- speed
- machinery location
- signals

Material handling systems

- automated
- non-automated

Moving stairs and walks

- capacity
- floors
- story height
- finish
- incline angle

- size
- number required
- speed
- machinery location
- special requirements

Pneumatic tube systems

- automatic
- size
- length

- manual
- stations
- special requirements

Vertical conveyer

- automatic

- non-automatic

The following illustrations show typical conveying systems used in interior projects. Figures 14.1a and 14.1b are schematic drawings of typical conveying systems. Figures 14.2a and 14.2b illustrate an elevator selective cost sheet from Means *Interior Cost Data*, 1987. This chart can be used to develop budget costs for various types of elevators, based on specific project requirements. Note the number of variables that can significantly affect the cost of elevators.

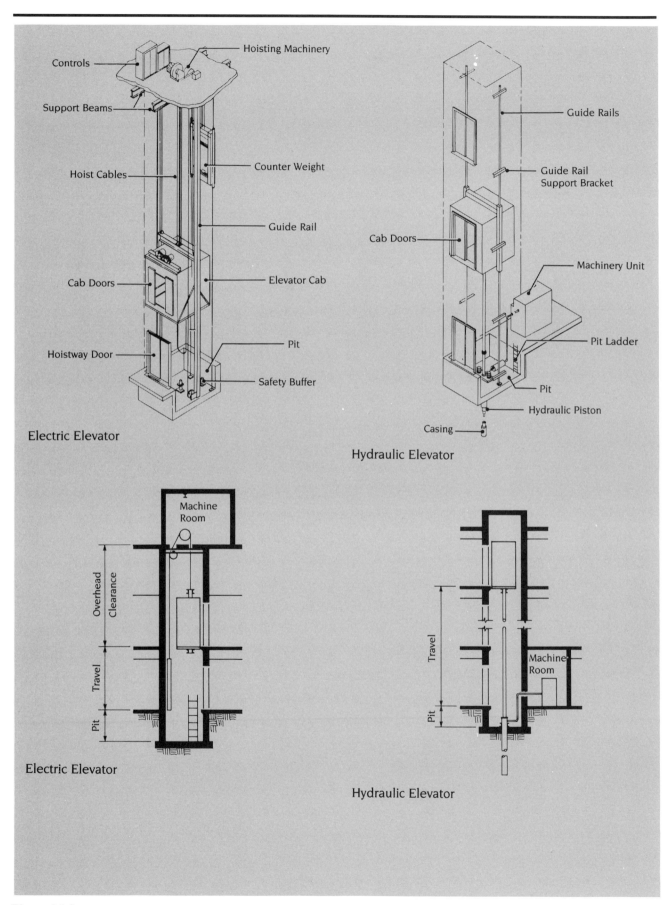

Controls

Hoisting Machinery

Support Beams

Hoist Cables

Counter Weight

Guide Rail

Cab Doors

Elevator Cab

Hoistway Door

Pit

Safety Buffer

Electric Elevator

Guide Rails

Guide Rail Support Bracket

Cab Doors

Machinery Unit

Pit Ladder

Pit

Hydraulic Piston

Casing

Hydraulic Elevator

Machine Room

Overhead Clearance

Travel

Pit

Electric Elevator

Travel

Machine Room

Pit

Hydraulic Elevator

Figure 14.1*a*

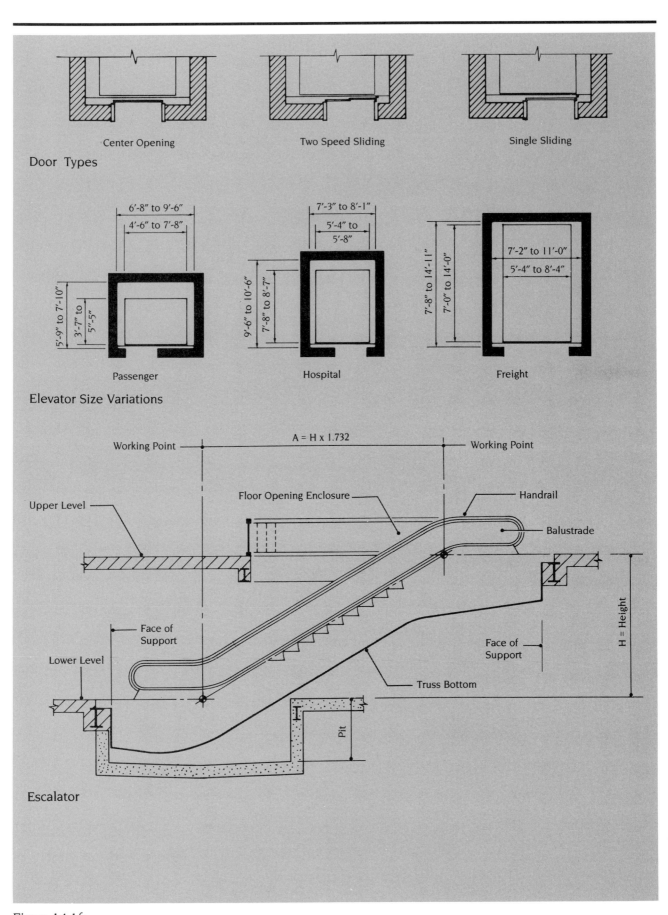

Door Types

Center Opening Two Speed Sliding Single Sliding

Elevator Size Variations

Passenger Hospital Freight

A = H x 1.732

Working Point Working Point

Upper Level Floor Opening Enclosure Handrail

Balustrade

Face of Support

Lower Level Face of Support

Truss Bottom

H = Height

Pit

Escalator

Figure 14.1*b*

CIRCLE REFERENCE NUMBERS

(123) Elevator Selective Costs (Div. 14.1-200)

	Passenger		Freight		Hospital	
A. Base Unit	Hydraulic	Electric	Hydraulic	Electric	Hydraulic	Electric
Capacity	1500 Lb.	2000 Lb.	2000 Lb.	4000 Lb.	3500 Lb.	3500 Lb.
Speed	50 F.P.M.	100 F.P.M.	25 F.P.M.	50 F.P.M.	50 F.P.M.	100 F.P.M.
#Stops/Travel Ft.	2/20	4/40	2/20	4/40	2/20	4/40
Push Button Oper.	Yes	Yes	Yes	Yes	Yes	Yes
Telephone Box & Wire	"	"	"	"	"	"
Emergency Lighting	"	"	No	No	"	"
Cab	Painted Steel	Painted Steel	Painted Steel	Painted Steel	S.S. Wainscot, Baked	Enamel Above
Cove Lighting	Yes	Yes	No	No	Yes	Yes
Floor	V.A.T.	V.A.T.	Wood w/Safety Treads	Wood w/Safety Treads	V.A.T.	V.A.T.
Doors, & Speedside Slide	Yes	Yes	No	No	Yes	Yes
Gates, Manual	No	No	Yes	Yes	No	No
Signals, Lighted Buttons	Car Only	Car Only	In Use Light	In Use Light	Car and Hall	Car and Hall
O.H. Geared Machine	N.A.	Yes	N.A.	Yes	N.A.	Yes
Variable Voltage Contr.	"	"	"	"	"	"
Emergency Alarm	"	"	"	"	"	"
Class "A" Loading	"	N.A.	Yes	"	"	N.A.
Base Cost	$37,500	$49,000	$29,900	$47,500	$41,700	$58,500
B. Capacity Adjustment						
2,000 Lb.	$ 2,450					
2,500	4,600	$ 2,350	$ 2,250			
3,000	5,200	2,800	3,600			
3,500	6,400	4,000	4,200			
4,000	6,800	4,850	5,400		$ 4,400	$ 2,925
4,500	8,000	5,700	5,800		5,050	3,350
5,000	9,400	7,000	7,200	$ 2,350	7,400	4,600
6,000			8,400	2,800		
7,000			9,900	4,000		
8,000			12,800	12,000		
10,000			15,000	15,000		
12,000			19,250	19,250		
16,000			22,500	22,500		
20,000			25,000	27,600		
C. Travel Over Base	$ 535 V.L.F.	$ 185 V.L.F.	$ 640 V.L.F.	$ 185 V.L.F.	$ 640 V.L.F.	$ 185 V.L.F.
D. Additional Stops	$ 5,025 Ea.	$ 5,025 Ea.	$ 4,300 Ea.	$ 4,300 Ea.	$ 5,200 Ea.	$ 5,025 Ea.
E. Speed Adjustment						
50 F.P.M.			$ 550			
75	$ 550		800	$ 1,875	$ 620	
100	1,000		1,500	2,975	1,010	
125	1,300		2,475	4,100	1,500	
150	3,575		3,400	5,250	2,450	
175				6,300	3,100	
Geared 200		$ 4,950		7,000		$ 5,800
4 Flrs. 250		6,950		8,200		7,950
Min. 300		8,550		10,200		9,550
350		9,800		11,700		10,600
400		10,700		12,800		11,700
Gearless 500		25,700		16,400		26,700
10 Flrs. 600		29,900		18,000		29,900
Min. 700		31,000		21,000		30,900
800		38,500		23,500		38,500
1,000		Spec. Applic.		Spec. Applic.		Spec. Applic.
1,200		"		"		"
F. Other Than Class						
"A" Loading						
"B"			$ 1,225	$ 1,225		
"C-1"			2,150	2,150		
"C-2"			1,350	1,350		
"C-3"			2,780	2,780		

251

Figure 14.2a

(123) Elevator Selective Costs (cont.)

	Passenger	Freight	Hospital
G. Options			
1. Controls			
Automatic, 2 car group	$ 3,500		$ 3,500
3 car group	6,100		6,100
4 car group	7,300		7,300
5 car group	9,100		9,100
6 car group	12,800		12,800
Emergency, fireman service	2,200		2,200
Intercom service	1,700		1,700
Selective collective, single car	2,500		2,500
Duplex car	3,850		3,850
2. Doors			
Center opening, 1 speed	$ 850		$ 850
2 speed	950		950
Rear opening-opposite front	—		5,350
Side opening, 2 speed	950		950
Freight, bi-parting	—	$3,000	—
Power operated door and gate	—	6,900	
3. Emergency power switching, automatic	$ 2,125		$ 2,125
Manual	1,075		1,075
4. Finishes based on 3500# cab			
Ceilings, acrylic panel	$ 225		—
Aluminum egg crate	275		$ 215
Doors, stainless steel	325		215
Floors, carpet, class "A"	80		—
Epoxy	215		215
Quarry tile	125		215
Slate	140		—
Steel plate	—	$ 535	—
Textured rubber	65		54
Walls, plastic laminate	300		215
Stainless steel	765		535
Return at door	430		430
Steel plate, 1/4" x 4' high, 14 ga. above	—	1,175	—
Entrance, doors, baked enamel	250		250
Stainless steel	430		430
Frames, baked enamel	250		250
Stainless steel	535		535
5. Maintenance contract - 12 months	$ 2,600	$1,650	$ 2,850
6. Signal devices			
Hall lantern, each	$ 375	$ 375	$ 375
Position indicator, car or lobby	270	270	·270
Add for over three each	65	65	65
7. Specialties			
High speed, heavy duty door opener	$ 550		$ 550
Variable voltage, O.H. gearless machine	26.700 - 53,400		26,700 - 53,400
Basement installed geared machine	6,000	$6,000	6,000

Figure 14.2b

Chapter 15
MECHANICAL

It is doubtful that an interiors estimator can spend the time or is capable of preparing a realistic unit price estimate for the mechanical trades. A reliable sub-bid should be sought for the mechanical portion of the interior work as soon as specific requirements are known. As a check, or for preliminary budgets, costs can be quickly calculated using the systems approach to mechanical estimating. A systems estimate is basically accomplished by counting the fixtures in the plumbing portion, establishing the class of fire protection required, and determining the heat and air conditioning load, source, and method of distribution. Appropriate systems costs are applied based on determined quantities and/or the size and occupancy use of the project.

For interior renovation projects, costs for the mechanical portion of the estimate must be estimated by using good judgment and past experience. In this type of work, the "cutting and patching" may be extensive, and existing conditions may severely restrict normal work procedures; both factors can significantly affect cost.

Mechanical work is usually broken down into three basic systems:
- Plumbing
- Fire protection
- Heating, Ventilation and Air Conditioning

Plumbing

In order to determine a good budget estimate for plumbing, the estimator must count the different fixture types and also determine other specific requirements. Systems costs can be applied by fixture as shown in Figures 15.1 and 15.2, from Means *Interior Cost Data* 1987. Note that listed in the "systems components" section are quantities for pipe and fittings as well as the fixture itself. These quantities (and costs) include rough-in material and labor to within ten feet of the waste and vent stacks, and hot and cold water supply piping.

Prior to or during design development, the estimator may be required to determine or anticipate plumbing fixture requirements for the proposed project. Figure 15.3, adapted from the BOCA Plumbing Code, can be used to determine fixture requirements at this stage. For final design purposes, local building officials must be consulted to assure compliance with applicable codes and requirements.

Costs for plumbing systems can also be determined by grouping, or by complete bathroom as shown in Figures 15.4 and 15.5 (From Means *Interior Cost Data*, 1987). Note that rough-in costs are included.

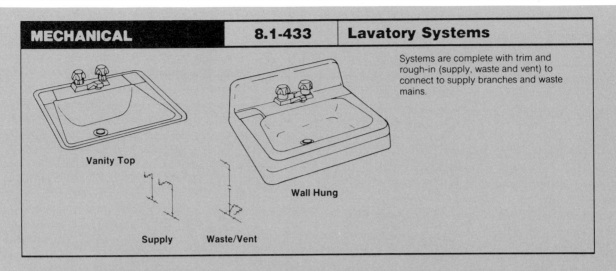

Systems are complete with trim and rough-in (supply, waste and vent) to connect to supply branches and waste mains.

Vanity Top

Wall Hung

Supply   Waste/Vent

System Components	QUANTITY	UNIT	COST EACH		
			MAT.	INST.	TOTAL
SYSTEM 08.1-433-1560					
LAVATORY W/TRIM, VANITY TOP, P.E. ON C.I., 20″ X 18″					
Lavatory w/trim, PE on CI, white, vanity top, 20″ x 18″ oval	1.000	Ea.	121	74	195
Pipe, steel, galvanized, schedule 40, threaded, 1-¼″ diam	4.000	L.F.	8.40	21.80	30.20
Copper tubing type DWV, solder joint, hanger 10′OC 1-¼″ diam	4.000	L.F.	6.84	35.76	42.60
Wrought copper DWV, Tee, sanitary, 1-¼″ diam	1.000	Ea.	3.45	29.55	33
P trap w/cleanout, 20 ga, 1-¼″ diam	1.000	Ea.	9.85	15.15	25
Copper tubing type L, solder joint, hanger 10′ OC ½″ diam	10.000	L.F.	7.80	50.70	58.50
Wrought copper 90° elbow for solder joints ½″ diam	2.000	Ea.	.36	26.84	27.20
Wrought copper Tee for solder joints, ½″ diam	2.000	Ea.	.66	41.34	42
Stop, chrome, angle supply, ½″ diam	2.000	Ea.	28.38	23.62	52
TOTAL			186.74	318.76	505.50

8.1-433	Lavatory Systems	COST EACH		
		MAT.	INST.	TOTAL
1560	Lavatory w/trim, vanity top, PE on CI, 20″ x 18″	185	320	505
1600	19″ x 16″ oval	180	325	505
1640	18″ round	175	320	495
1680	Cultured marble, 19″ x 17″	145	325	470
1720	25″ x 19″	160	320	480
1760	Stainless, self-rimming, 25″ x 22″	205	320	525
1800	17″ x 22″	195	320	515
1840	Steel enameled, 20″ x 17″	135	325	460
1880	19″ round	130	330	460
1920	Vitreous china, 20″ x 16″	245	335	580
1960	19″ x 16″	190	335	525
2000	22″ x 13″	195	335	530
2040	Wall hung, PE on CI, 18″ x 15″	345	350	695
2080	19″ x 17″	310	350	660
2120	20″ x 18″	260	350	610
2160	Vitreous china, 18″ x 15″	255	360	615
2200	19″ x 17″	205	360	565
2240	24″ x 20″	310	355	665

317

Figure 15.1

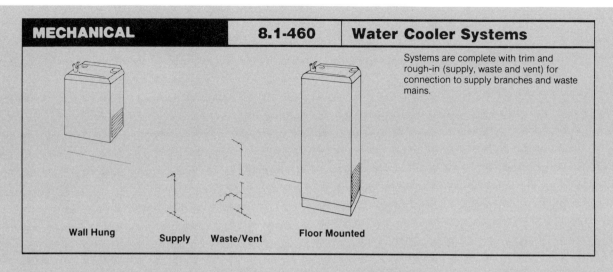

Systems are complete with trim and rough-in (supply, waste and vent) for connection to supply branches and waste mains.

Wall Hung   **Supply**   **Waste/Vent**   **Floor Mounted**

System Components	QUANTITY	UNIT	COST EACH		
			MAT.	INST.	TOTAL
SYSTEM 08.1-460-1840					
WATER COOLER, ELECTRIC, SELF CONTAINED, WALL HUNG, 8.2 GPH					
Water cooler, wall mounted, 8.2 GPH	1.000	Ea.	368.50	121.50	490
Copper tubing type DWV, solder joint, hanger 10'OC 1-¼" diam	4.000	L.F.	6.84	35.76	42.60
Wrought copper DWV, Tee, sanitary 1-¼" diam	1.000	Ea.	3.45	29.55	33
P trap, copper drainage, 1-¼" diam	1.000	Ea.	9.85	15.15	25
Copper tubing type L, solder joint, hanger 10' OC ⅜" diam	5.000	L.F.	3.30	23.95	27.25
Wrought copper 90° elbow for solder joints ⅜" diam	1.000	Ea.	.48	12.22	12.70
Wrought copper Tee for solder joints, ⅜" diam	1.000	Ea.	1.18	18.82	20
Stop and waste, straightway, bronze, solder, ⅜" diam	1.000	Ea.	3.96	11.19	15.15
TOTAL			397.56	268.14	665.70

8.1-460	Water Cooler Systems	COST EACH		
		MAT.	INST.	TOTAL
1840	Water cooler, electric, wall hung, 8.2 GPH	400	270	670
1880	Dual height, 14.3 GPH	555	275	830
1920	Wheelchair type, 7.5 G.P.H.	980	270	1,250
1960	Semi recessed, 8.1 G.P.H.	515	270	785
2000	Full recessed, 8 G.P.H.	840	285	1,125
2040	Floor mounted, 14.3 G.P.H.	420	235	655
2080	Dual height, 14.3 G.P.H.	590	280	870
2120	Refrigerated compartment type, 1.5 G.P.H.	760	235	995
2160	Cafeteria type, dual glass fillers, 27 G.P.H.	1,875	335	2,210

320

Figure 15.2

CIRCLE REFERENCE NUMBERS

(132) **Minimum Plumbing Fixture Requirements**

TYPE OF BUILDING/USE	WATER CLOSETS		URINALS		LAVATORIES		BATHTUBS OR SHOWERS		DRINKING FOUNTAIN	OTHER
	Persons	Fixtures	Persons	Fixtures	Persons	Fixtures	Persons	Fixtures	Fixtures	Fixtures
Assembly Halls Auditoriums Theater Public assembly	1-100 101-200 201-400	1 2 3	1-200 201-400 401-600	1 2 3	1-200 201-400 401-750	1 2 3			1 for each 1000 persons	1 service sink
	Over 400 add 1 fixt. for ea. 500 men: 1 fixt. for ea. 300 women		Over 600 add 1 fixture for each 300 men		Over 750 add 1 fixture for each 500 persons					
Assembly Public Worship	300 men 150 women	1 1	300 men	1	men women	1 1			1	
Dormitories	Men: 1 for each 10 persons Women: 1 for each 8 persons		1 for each 25 men, over 150 add 1 fixture for each 50 men		1 for ea. 12 persons 1 separate dental lav. for each 50 persons recom.		1 for ea. 8 persons For women add 1 additional for each 30. Over 150 persons add 1 for each 20.		1 for each 75 persons	Laundry trays 1 for each 50 serv. sink 1 for ea. 100
Dwellings Apartments and homes	1 fixture for each unit				1 fixture for each unit		1 fixture for each unit			
Hospitals Room Ward Waiting room	Ea. 8 persons	1			Ea. 10 persons	1		1	1 for 100 patients	1 service sink per floor
Industrial Mfg. plants Warehouses	1-10 11-25 26-50 51-75 76-100	1 2 3 4 5	0-30 31-80 81-160 161-240	1 2 3 4	1-100 over 100	1 for ea. 10 1 for ea. 15	1 Shower for each 15 persons subject to excessive heat or occupation-al hazard		1 for each 75 person	
	1 fixture for each additional 30 persons									
Public Buildings Businesses Offices	1-15 16-35 36-55 56-80 81-110 111-150	1 2 3 4 5 6	Urinals may be provided in place of water closets but may not replace more than 1/3 required number of men's water closets		1-15 16-35 36-60 61-90 91-125	1 2 3 4 5			1 for each 75 persons	1 service sink per floor
	1 fixture for ea. additional 40 persons				1 fixture for ea. additional 45 persons					
Schools Elementary	1 for ea. 30 boys 1 for ea. 25 girls		1 for ea. 25 boys		1 for ea. 35 boys 1 for ea. 35 girls		For gym or pool shower room 1/5 of a class		1 for each 40 pupils	
Schools Secondary	1 for ea. 40 boys 1 for ea. 30 girls		1 for ea. 25 boys		1 for ea. 40 boys 1 for ea. 40 girls		For gym or pool shower room 1/5 of a class		1 for each 50 pupils	

Figure 15.3

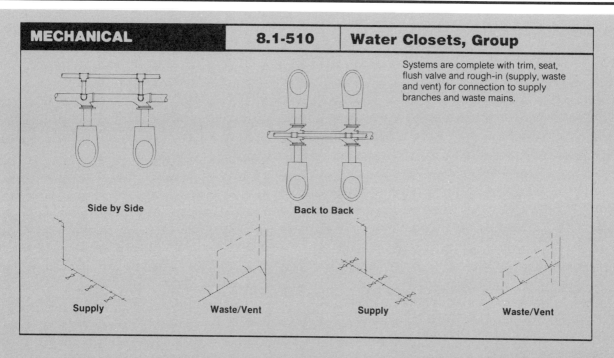

MECHANICAL	8.1-510	Water Closets, Group

Systems are complete with trim, seat, flush valve and rough-in (supply, waste and vent) for connection to supply branches and waste mains.

Side by Side

Back to Back

Supply **Waste/Vent** **Supply** **Waste/Vent**

System Components	QUANTITY	UNIT	COST EACH		
			MAT.	INST.	TOTAL
SYSTEM 08.1-510-1760					
WATER CLOSETS, BATTERY MOUNT, WALL HUNG, SIDE BY SIDE, FIRST CLOSET					
Water closet, bowl only w/flush valve, seat, wall hung	1.000	Ea.	258.50	81.50	340
Pipe, CI soil, no hub, cplg 10' OC, hanger 5' OC, 4" diam	3.000	L.F.	15.60	27.30	42.90
Coupling, standard, CI, soil, no hub, 4" diam	2.000	Ea.	6.82	43.18	50
Copper tubing type L, solder joints, hangers 10' OC, 1" diam	6.000	L.F.	9.66	44.64	54.30
Copper tubing, type DWV, solder joints, hangers 10'OC, 2" diam	6.000	L.F.	16.44	66.96	83.40
Wrought copper 90° elbow for solder joints 1" diam	1.000	Ea.	.99	16.76	17.75
Wrought copper Tee for solder joints, 1" diam	1.000	Ea.	2.68	27.32	30
Support/carrier, siphon jet, horiz, adjustable single, 4" pipe	1.000	Ea.	97.90	47.10	145
Valve, gate, bronze, 125 lb, NRS, soldered 1" diam	1.000	Ea.	19.91	14.09	34
Wrought copper, DWV, 90° elbow, 2" diam	1.000	Ea.	3.99	27.01	31
TOTAL			432.49	395.86	828.35

8.1-510	Water Closets, Group	COST EACH		
		MAT.	INST.	TOTAL
1760	Water closets, battery mount, wall hung, side by side, first closet	430	395	825
1800	Each additional water closet, add	415	375	790
3000	Back to back, first pair of closets	780	520	1,300
3100	Each additional pair of closets, back to back	760	505	1,265

322

Figure 15.4

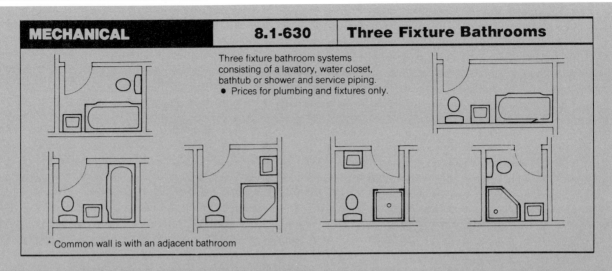

MECHANICAL	8.1-630	Three Fixture Bathrooms

Three fixture bathroom systems consisting of a lavatory, water closet, bathtub or shower and service piping.
● Prices for plumbing and fixtures only.

* Common wall is with an adjacent bathroom

System Components	QUANTITY	UNIT	COST EACH		
			MAT.	INST.	TOTAL
SYSTEM 08.1-630-1170					
BATHROOM, LAVATORY, WATER CLOSET & BATHTUB					
ONE WALL PLUMBING, STAND ALONE					
Wtr closet, 2 pc close cpld vit china flr mntd w/seat supply & stop	1.000	Ea.	126.50	93.50	220
Water closet, rough-in waste & vent	1.000	Set	90.97	249.03	340
Lavatory w/ftngs, wall hung, white, PE on CI, 20" x 18"	1.000	Ea.	137.50	62.50	200
Lavatory, rough-in waste & vent	1.000	Set	122.21	292.79	415
Bathtub, white PE on CI, w/ftgs, mat bottom, recessed, 5' long	1.000	Ea.	280.50	109.50	390
Baths, rough-in waste and vent	1.000	Set	69.16	250.34	319.50
TOTAL			826.84	1,057.66	1,884.50

8.1-630	Three Fixture Bathroom, One Wall Plumbing	COST EACH		
		MAT.	INST.	TOTAL
1150	Bathroom, three fixture, one wall plumbing			
1160	Lavatory, water closet & bathtub			
1170	Stand alone	825	1,050	1,875
1180	Share common plumbing wall *	720	800	1,520

8.1-630	Three Fixture Bathroom, Two Wall Plumbing	COST EACH		
		MAT.	INST.	TOTAL
2130	Bathroom, three fixture, two wall plumbing			
2140	Lavatory, water closet & bathtub			
2160	Stand alone	830	1,075	1,905
2180	Long plumbing wall common *	750	880	1,630
3610	Lavatory, bathtub & water closet			
3620	Stand alone	885	1,225	2,110
3640	Long plumbing wall common *	840	1,100	1,940
4660	Water closet, corner bathtub & lavatory			
4680	Stand alone	1,500	1,075	2,575
4700	Long plumbing wall common *	1,425	800	2,225
6100	Water closet, stall shower & lavatory			
6120	Stand alone	895	1,375	2,270
6140	Long plumbing wall common *	855	1,275	2,130
7060	Lavatory, corner stall shower & water closet			
7080	Stand alone	1,050	1,250	2,300
7100	Short plumbing wall common *	980	890	1,870

324

Figure 15.5

For most interior projects, plumbing stacks are provided to the space. Toilet rooms and other fixtures are usually located close to this piping. However, when the plumbing fixtures are located away from the stacks or if the stacks are to be part of the work being estimated, costs must be determined and included. For small quantities, costs for pipe (linear feet) and fittings (each) can be included. For budgeting large projects, percentage multipliers, such as those shown in Figure 15.6 can be useful.

Fire Protection

Square foot costs for fire protection systems should be developed from past projects for budget purposes. These costs may be based on the relative hazard of occupancy — light, ordinary and extra — and on the system type. A comparison of some requirements of the different hazards is shown in Figure 15.7. Consideration must also be given to special or unusual requirements. For example, many architects specify that sprinkler heads must be located in the center of ceiling tiles. Each head may require extra elbows and nipples for precise location. Recessed heads are more expensive. Special dry pendent heads are required in areas subject to freezing. When installing a sprinkler system in an existing structure, a completely new water service may be required in addition to the existing domestic water service. These are just a few examples of requirements which may necessitate an adjustment of square foot costs.

In addition to the hazard, the *type* of sprinkler system is the most significant factor affecting cost. The size of system (square footage) and the number of floors served by one system will also affect the cost. This cost variation is illustrated along with the components of a typical wet pipe sprinkler system in Figures 15.8a and 15.8b.

Wet Pipe Systems employ automatic sprinklers attached to a piping system containing water and connected to a water supply so that water discharges immediately from sprinklers opened by a fire.

Plumbing Approximations for Quick Estimating

Water Control
Water Meter; Backflow Preventer;
Shock Absorbers; Vacuum Breakers; .... 10 to 15% of Fixtures
Mixer.

Pipe And Fittings: .... 30 to 60% of Fixtures

    Note: Lower percentage for compact buildings or larger buildings with plumbing in one area.
            Larger percentage for large buildings with plumbing spread out.
            In extreme cases pipe may be more than 100% of fixtures.
            Percentages **do not** include special purpose or process piping.

Plumbing Labor:
1 & 2 Story Residential .... Rough-in Labor = 80% of Materials
Apartment Buildings .... Rough-in Labor = 90 to 100% of Materials
Labor for handling and placing fixtures is approximately 25 to 30% of fixtures.

Quality/Complexity Multiplier (For all installations)
Economy installation, add .... 0 to 5%
Good quality, medium complexity, add .... 5 to 15%
Above average quality and complexity, add .... 15 to 25%

Figure 15.6

Sprinkler System Classification

Rules for installation of sprinkler systems vary depending on the classification of occupancy falling into one of three categories as follows:

Light Hazard Occupancy	Ordinary Hazard Occupancy	Extra Hazard Occupancy
The protection area allotted per sprinkler should not exceed 200 S.F. with the maximum distance between lines and sprinklers on lines being 15'. The sprinklers do not need to be staggered. Branch lines should not exceed eight sprinklers on either side of a cross main. Each large area requiring more than 100 sprinklers and without a sub-dividing partition should be supplied by feed mains or risers sized for ordinary hazard occupancy.	The protection area allotted per sprinkler shall not exceed 130 S.F. of noncombustible ceiling and 120 S.F. of combustible ceiling. The maximum allowable distance between sprinkler lines and sprinklers on line is 15'. Sprinklers shall be staggered if the distance between heads exceed 12'. Branch lines should not exceed eight sprinklers on either side of a cross main.	The protection area allotted per sprinkler shall not exceed 90 S.F. of noncombustible ceiling and 80 S.F. of combustible ceiling. The maximum allowable distance between lines and between sprinklers on lines is 12'. Sprinklers on alternate lines shall be staggered if the distance between sprinklers on lines exceeds 8'. Branch lines should not exceed six sprinklers on either side of a cross main.
Included in this group are:	Included in this group are:	Included in this group are:

Light Hazard		Ordinary Hazard		Extra Hazard	
Auditoriums	Museums	Automotive garages	Electric generating	Aircraft hangars	Paint shops
Churches	Nursing Homes	Bakeries	stations	Chemical works	Shade cloth
Clubs	Offices	Beverage	Feed mills	Explosives	manufacturing
Educational	Residential	manufacturing	Grain elevators	manufacturing	Solvent extracting
Hospitals	Restaurants	Bleacheries	Ice manufacturing	Linseed	Varnish works
Institutional	Schools	Boiler houses	Laundries	manufacturing	Volatile flammable
Libraries	Theaters	Canneries	Machine shops	Linseed oil mills	liquid manufacturing
(except large stack rooms)		Cement plants	Mercantiles	Oil refineries	& use
		Clothing factories	Paper mills		
		Cold storage	Printing and		
		warehouses	publishing		
		Dairy products	Shoe factories		
		manufacturing	Warehouses		
		Distilleries	Wood product		
		Dry cleaning	assembly		

Figure 15.7

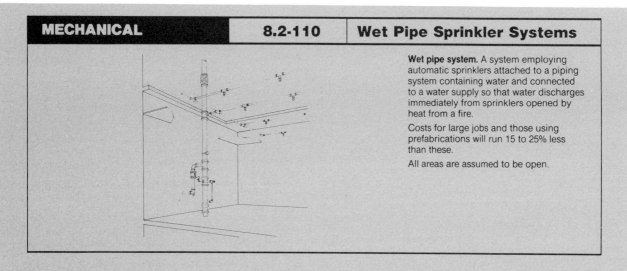

MECHANICAL | 8.2-110 | Wet Pipe Sprinkler Systems

Wet pipe system. A system employing automatic sprinklers attached to a piping system containing water and connected to a water supply so that water discharges immediately from sprinklers opened by heat from a fire.

Costs for large jobs and those using prefabrications will run 15 to 25% less than these.

All areas are assumed to be open.

System Components	QUANTITY	UNIT	COST EACH		
			MAT.	INST.	TOTAL
SYSTEM 08.2-110-0580					
WET PIPE SPRINKLER, STEEL, BLACK, SCH. 40 PIPE					
LIGHT HAZARD, ONE FLOOR, 2000 S.F.					
Valve, gate, iron body, 125 lb, OS&Y, flanged, 4" diam	1.000	Ea.	148.50	121.50	270
Valve, swing check, bronze, 125 lb, regrinding disc, 2-½" pipe size	1.000	Ea.	64.35	25.65	90
Valve, angle, bronze, 150 lb, rising stem, threaded, 2" diam	1.000	Ea.	80.03	17.48	97.51
*Alarm valve, 2-½" pipe size	1.000	Ea.	400.13	124.88	525.01
Alarm, water motor, complete with gong	1.000	Ea.	79.61	51.64	131.25
Valve, swing check, w/balldrip CI with brass trim 4" pipe size	1.000	Ea.	84.98	125.03	210.01
Pipe, steel, black, schedule 40, 4" diam	10.000	L.F.	56.78	100.73	157.51
*Flow control valve, trim & gauges, 4" pipe size	1.000	Set	787.88	280.88	1,068.76
Fire alarm horn, electric	1.000	Ea.	21.45	29.55	51
Pipe, steel, black, schedule 40, threaded, cplg & hngr 10'OC, 2-½" diam	20.000	L.F.	74.85	144.90	219.75
Pipe, steel, black, schedule 40, threaded, cplg & hngr 10'OC, 2" diam	12.500	L.F.	28.50	70.88	99.38
Pipe, steel, black, schedule 40, threaded, cplg & hngr 10'OC, 1-¼" diam	37.500	L.F.	53.72	153	206.72
Pipe steel, black, schedule 40, threaded cplg & hngr 10'OC, 1" diam	112.000	L.F.	128.52	425.88	554.40
Pipe Tee, malleable iron black, 150 lb threaded, 4" pipe size	2.000	Ea.	57.75	182.25	240
Pipe Tee, malleable iron black, 150 lb threaded, 2-½" pipe size	2.000	Ea.	22.28	81.23	103.51
Pipe Tee, malleable iron black, 150 lb threaded, 2" pipe size	1.000	Ea.	5	33.26	38.26
Pipe Tee, malleable iron black, 150 lb threaded, 1-¼" pipe size	5.000	Ea.	13.84	128.66	142.50
Pipe Tee, malleable iron black, 150 lb threaded, 1" pipe size	4.000	Ea.	6.81	101.19	108
Pipe 90° elbow, malleable iron black, 150 lb threaded, 1" pipe size	6.000	Ea.	6.57	92.43	99
Sprinkler head, standard spray, brass 135°-286°F ½" NPT, ⅜" orifice	12.000	Ea.	27.27	251.73	279
Valve, gate, bronze, NRS, class 150, threaded, 1" pipe size	1.000	Ea.	13.53	10.47	24
*Standpipe connection, wall, single, flush w/plug & chain 2-½"x2-½"	1.000	Ea.	70.95	75.30	146.25
TOTAL			2,233.30	2,628.52	4,861.82
COST PER S.F.			1.12	1.31	2.43

*Not included in systems under 2000 S.F.

8.2-110	Wet Pipe Sprinkler Systems	COST PER S.F.		
		MAT.	INST.	TOTAL
0520	Wet pipe sprinkler systems, steel, black, sch. 40 pipe			
0530	Light hazard, one floor, 500 S.F.	.70	1.33	2.03
0560	1000 S.F.	1.03	1.37	2.40
0580	2000 S.F.	1.12	1.31	2.43
0600	5000 S.F.	.55	.96	1.51
0620	10,000 S.F.	.37	.81	1.18

327

Figure 15.8a

8.2-110	Wet Pipe Sprinkler Systems	COST PER S.F.		
		MAT.	INST.	TOTAL
0640	50,000 S.F.	.29	.73	1.02
0660	Each additional floor, 500 S.F.	.40	1.11	1.51
0680	1000 S.F.	.36	1.02	1.38
0700	2000 S.F.	.32	.90	1.22
0720	5000 S.F.	.23	.79	1.02
0740	10,000 S.F.	.23	.73	.96
0760	50,000 S.F.	.25	.69	.94
1000	Ordinary hazard, one floor, 500 S.F.	.79	1.46	2.25
1020	1000 S.F.	1.02	1.32	2.34
1040	2000 S.F.	1.19	1.42	2.61
1060	5000 S.F.	.63	1.04	1.67
1080	10,000 S.F.	.47	1.07	1.54
1100	50,000 S.F.	.45	1.04	1.49
1140	Each additional floor, 500 S.F.	.50	1.25	1.75
1160	1000 S.F.	.34	1	1.34
1180	2000 S.F.	.39	1	1.39
1200	5000 S.F.	.40	.96	1.36
1220	10,000 S.F.	.33	.99	1.32
1240	50,000 S.F.	.31	.93	1.24
1500	Extra hazard, one floor, 500 S.F.	2.74	2.43	5.17
1520	1000 S.F.	1.66	1.99	3.65
1540	2000 S.F.	1.23	1.85	3.08
1560	5000 S.F.	.84	1.66	2.50
1580	10,000 S.F.	.81	1.53	2.34
1600	50,000 S.F.	.83	1.45	2.28
1660	Each additional floor, 500 S.F.	.59	1.51	2.10
1680	1000 S.F.	.58	1.44	2.02
1700	2000 S.F.	.49	1.44	1.93
1720	5000 S.F.	.41	1.29	1.70
1740	10,000 S.F.	.46	1.17	1.63
1760	50,000 S.F.	.44	1.09	1.53
2020	Grooved steel, black sch. 40 pipe, light hazard, one floor, 2000 S.F.	1.20	1.15	2.35
2060	10,000 S.F.	.50	.73	1.23
2100	Each additional floor, 2000 S.F.	.40	.73	1.13
2150	10,000 S.F.	.27	.62	.89
2200	Ordinary hazard, one floor, 2000 S.F.	1.22	1.23	2.45
2250	10,000 S.F.	.50	.91	1.41
2300	Each additional floor, 2000 S.F.	.42	.82	1.24
2350	10,000 S.F.	.36	.83	1.19
2400	Extra hazard, one floor, 2000 S.F.	1.34	1.59	2.93
2450	10,000 S.F.	.73	1.20	1.93
2500	Each additional floor, 2000 S.F.	.61	1.19	1.80
2550	10,000 S.F.	.50	1.07	1.57
3050	Grooved steel black sch. 10 pipe, light hazard, one floor, 2000 S.F.	1.20	1.14	2.34
3100	10,000 S.F.	.43	.69	1.12
3150	Each additional floor, 2000 S.F.	.40	.72	1.12
3200	10,000 S.F.	.28	.61	.89
3250	Ordinary hazard, one floor, 2000 S.F.	1.22	1.22	2.44
3300	10,000 S.F.	.51	.89	1.40
3350	Each additional floor, 2000 S.F.	.42	.81	1.23
3400	10,000 S.F.	.37	.81	1.18
3450	Extra hazard, one floor, 2000 S.F.	1.34	1.58	2.92
3500	10,000 S.F.	.68	1.18	1.86
3550	Each additional floor, 2000 S.F.	.61	1.18	1.79
3600	10,000 S.F.	.49	1.06	1.55
4050	Copper tubing, type M, light hazard, one floor, 2000 S.F.	1.11	1.43	2.54
4100	10,000 S.F.	.39	.99	1.38
4150	Each additional floor, 2000 S.F.	.32	1.03	1.35

328

Figure 15.8*b*

Dry Pipe Systems employ automatic sprinklers attached to a piping system containing air under pressure, the release of which, as from the opening of sprinklers, permits the water pressure to open a valve known as a "dry pipe valve". The water then flows into the piping system and out the opened sprinklers.

Pre-Action Systems employ automatic sprinklers attached to a piping system containing air that may or may not be under pressure, with a supplemental heat responsive system of generally more sensitive characteristics than the automatic sprinklers themselves, installed in the same areas as the sprinklers. Actuation of the heat responsive system, as from a fire, opens a valve which permits water to fill the sprinkler piping system and to be discharged from any sprinklers which may open.

Deluge Systems employ open sprinklers attached to a piping system, connected to a water supply through a valve which is opened by the operation of a heat responsive system installed in the same areas as the sprinklers. When this valve opens, water flows into the piping system and discharges from all sprinklers at once.

Combined Dry Pipe and Pre-Action Sprinkler Systems employ automatic sprinklers attached to a piping system containing air under pressure, with a supplemental heat responsive system. This supplemental system is generally more sensitive than the automatic sprinklers themselves, and is installed in the same area as the sprinklers. Operation of the heat responsive system also opens approved air exhaust valves at the end of the feed main. This facilitates the filling of the system with water, which usually precedes the opening of sprinklers. The heat responsive system also serves as an automatic fire alarm system.

Limited Water Supply Systems employ automatic sprinklers and conforming to these standards but supplied by a pressure tank of limited capacity.

Chemical Systems use halon, carbon dioxide, dry chemicals, or high expansion foam as selected for special requirements. The chemical agent may extinguish flames by excluding oxygen, interrupting chemical action of the oxygen uniting with fuel, or sealing and cooling the combustion center.

Firecycle Systems are fixed fire protection sprinkler systems utilizing water as its extinguishing agent. These are time delayed, recycling, preaction type systems which automatically shut the water off when heat is reduced below the detector operating temperature and turn the water back on when that temperature is exceeded. The system senses a fire condition through a closed circuit electrical detector system which controls water flow to the fire automatically. Batteries supply up to a 90 hour emergency power supply for system operation. The piping system is dry (until water is required) and is monitored with pressurized air. Should any leak in the system piping occur, an alarm will sound, but water will not enter the system until heat is sensed by a firecycle detector.

Heating, Ventilating, and Air Conditioning

As with fire protection, square foot (or systems) costs can be developed for HVAC (heating, ventilating, and air conditioning) by keeping records from past projects. It is recommended that the estimator obtain quotations or detailed estimates from experienced engineers or subcontractors whenever possible. HVAC is a very specialized trade that requires specific knowledge to estimate properly. However, budgets can be developed based on square foot, cubic foot, or systems costs. Such costs will vary based on the type of system and the use of the occupied space as well as the size of the space. Different types of systems are illustrated in Figures 15.9a and 9b. Costs for each of these types of systems are developed in Means *Interior Cost Data*. An example for air-cooled split systems is shown in Figure 15.10. (Note: Heating would be provided from another source, but could use the air handling unit and ductwork for distribution.)

For preliminary budgets, prior to or during design development, the estimator can determine rough heating and cooling requirements in order to determine costs. The tables in Figure 15.11 can be used to determine heat loss (in BTU's per hour). This figure, for all practical purposes, is the capacity required of the heat source (e.g. boiler, furnace). Figure 15.12 can be used to determine cooling requirements in tons for 45 types of building uses. (One ton of cooling equals 12,000 BTU's per hour.) When heating and cooling systems are combined, as with rooftop systems, the cooling requirements are used to determine the size of the system. When the system capacity has been determined, budget costs can be obtained based on building use as shown in Figure 15.10. For systems and spaces other than the capacities and sizes shown, costs may be interpolated.

For most interiors projects, the main heating and cooling systems will be in place. Distribution (ductwork, hot water baseboard, fan coil units, etc.), however, is usually dependent upon final design and is often executed as part of the interior project. Computer rooms usually require complete, often independent, cooling systems as part of the interior work. (Computer rooms rarely require supplemental heating.)

Computer rooms impose special requirements on air conditioning systems. A prime requirement is reliability, due to the potential monetary loss that could be incurred by a system failure. A second basic requirement is the tolerance of control with which temperature and humidity are regulated, and dust eliminated. Because the air conditioning system reliability is so vital, the additional cost of reserve capacity and redundant components is often justified.

Computer areas may be environmentally controlled by one of three methods as follows:

- *Self-contained Units*. These are units built to higher standards of performance and reliability, and usually contain alarms and controls to indicate component operation failure, filter change necessity, etc. It should be remembered that these units in the room will occupy space that is relatively expensive to build and that all alterations and service of the equipment will also have to be accomplished within the computer area.
- *Decentralized Air Handling Units*. In operation, this type is similar to the self-contained units except that cooling capability comes from remotely located refrigeration equipment as refrigerant or chilled water. Since no compressors or refrigerating equipment are required in the air units, they are smaller and require less service than self-contained units. This type of system may also be beneficial if some of the computer components require chilled water for cooling.
- *Central System Supply*. Cooling is obtained from a central source which, since it is not located within the computer room, may have excess capacity and permit greater flexibility without interfering with the computer components. System performance criteria must still be met. This type of system may provide less control than independent units.

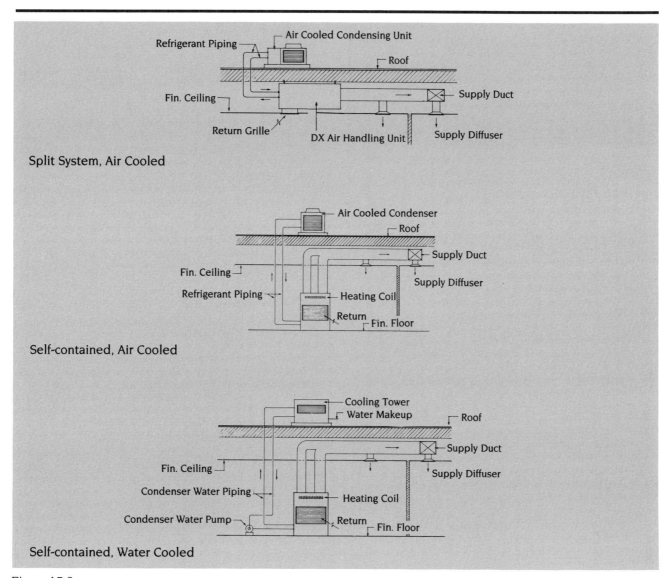

Split System, Air Cooled

Self-contained, Air Cooled

Self-contained, Water Cooled

Figure 15.9a

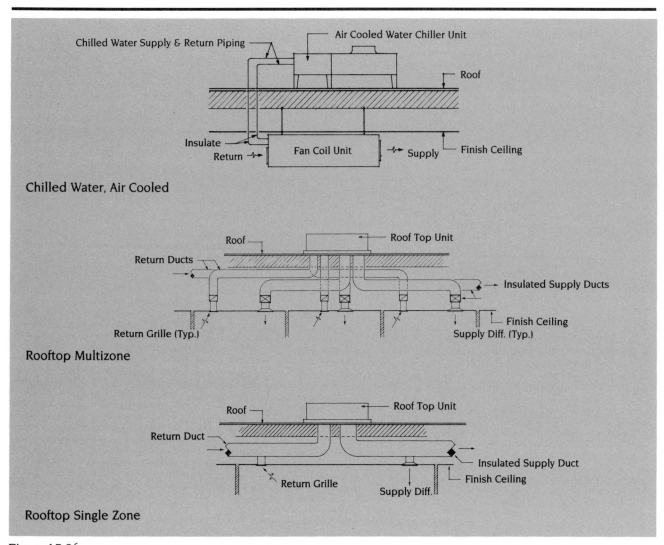

Chilled Water, Air Cooled

Rooftop Multizone

Rooftop Single Zone

Figure 15.9b

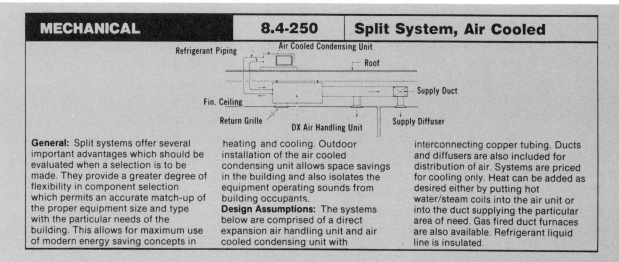

Refrigerant Piping — Air Cooled Condensing Unit — Roof — Supply Duct — Fin. Ceiling — Return Grille — DX Air Handling Unit — Supply Diffuser

General: Split systems offer several important advantages which should be evaluated when a selection is to be made. They provide a greater degree of flexibility in component selection which permits an accurate match-up of the proper equipment size and type with the particular needs of the building. This allows for maximum use of modern energy saving concepts in heating and cooling. Outdoor installation of the air cooled condensing unit allows space savings in the building and also isolates the equipment operating sounds from building occupants.

Design Assumptions: The systems below are comprised of a direct expansion air handling unit and air cooled condensing unit with interconnecting copper tubing. Ducts and diffusers are also included for distribution of air. Systems are priced for cooling only. Heat can be added as desired either by putting hot water/steam coils into the air unit or into the duct supplying the particular area of need. Gas fired duct furnaces are also available. Refrigerant liquid line is insulated.

System Components	QUANTITY	UNIT	COST EACH		
			MAT.	INST.	TOTAL
SYSTEM 08.4-250-1280					
SPLIT SYSTEM, AIR COOLED CONDENSING UNIT					
APARTMENT CORRIDORS, 1,000 S.F., 1.80 TON					
Fan coil AC unit, cabinet mntd & filters direct expansion air cool	1.000	Ea.	245.59	60.02	305.61
Ductwork package, for split system, remote condensing unit	1.000	System	248.01	342.16	590.17
Refrigeration piping	1.000	System	70.53	368.67	439.20
Condensing unit, air cooled, incls compressor & standard controls	1.000	Ea.	650.87	239.73	890.60
TOTAL			1,215	1,010.58	2,225.58
COST PER S.F.			1.22	1.01	2.23

*Cooling requirements would lead to more than one system

8.4-250	Split Systems With Air Cooled Condensing Units	COST PER S.F.		
		MAT.	INST.	TOTAL
1260	Split system, air cooled condensing unit			
1280	Apartment corridors, 1,000 S.F., 1.83 ton	1.22	1.01	2.23
1440	20,000 S.F., 36.66 ton	1.46	1.43	2.89
1520	Banks and libraries, 1,000 S.F., 4.17 ton	2.51	2.33	4.84
1680	20,000 S.F., 83.32 ton	3.84	3.39	7.23
1760	Bars and taverns, 1,000 S.F., 11.08 ton	6.90	4.60	11.50
1880	10,000 S.F., 110.84 ton	8.40	5.35	13.75
2000	Bowling alleys, 1,000 S.F., 5.66 ton	4.18	4.38	8.56
2160	20,000 S.F., 113.32 ton	5.45	4.79	10.24
2320	Department stores, 1,000 S.F., 2.92 ton	1.76	1.60	3.36
2480	20,000 S.F., 58.33 ton	2.32	2.28	4.60
2560	Drug stores, 1,000 S.F., 6.66 ton	4.92	5.15	10.07
2720	20,000 S.F., 133.32 ton*			
2800	Factories, 1,000 S.F., 3.33 ton	2.01	1.83	3.84
2960	20,000 S.F., 66.66 ton	3.05	2.73	5.78
3040	Food supermarkets, 1,000 S.F., 2.83 ton	1.71	1.56	8.27
3200	20,000 S.F., 56.66 ton	2.25	2.21	4.46
3280	Medical centers, 1,000 S.F., 2.33 ton	1.43	1.25	2.68
3440	20,000 S.F., 46.66 ton	1.86	1.82	3.68
3520	Offices, 1,000 S.F., 3.17 ton	1.92	1.74	3.66
3680	20,000 S.F., 63.32 ton	2.90	2.60	5.50
3760	Restaurants, 1,000 S.F., 5.00 ton	3.69	3.87	7.56
3920	20,000 S.F., 100.00 ton	4.80	4.22	9.02
4000	Schools and colleges, 1,000 S.F., 3.83 ton	2.31	2.14	4.45
4160	20,000 S.F., 76.66 ton	3.51	3.13	6.64

343

Figure 15.10

Factors for Determining Heat Loss for Various Types of Buildings

General: While the most accurate estimates of heating requirements would naturally be based on detailed information about the building being considered, it is possible to arrive at a reasonable approximation using the following procedure:

1. Calculate the cubic volume of the room or building.
2. Select the appropriate factor from Table A. Note that the factors apply only to inside temperatures listed in the first column and to 0°F outside temperature.
3. If the building has bad north and west exposures, multiply the heat loss factor by 1.1
4. If the outside design temperature is other than 0°F, multiply the factor from Table A by the factor from Table B.
5. Multiply the cubic volume by the factor selected from Table A. This will give the estimated BTU per hour heat loss which must be made up to maintain inside temperature.

Table A
Factories & Industrial Plants General Office Area 70°F

Conditions	Qualifications	Loss Factor*
One Story	Skylight in Roof	6.2
	No Skylight in Roof	5.7
Multiple Story	Two Story	4.6
	Three Story	4.3
	Four Story	4.1
	Five Story	3.9
	Six Story	3.6
All Walls Exposed	Flat Roof	6.9
	Heated Space Above	5.2
One Long Warm Common Wall	Flat Roof	6.3
	Heated Space Above	4.7
Warm Common Walls on Both Long Sides	Flat Roof	5.8
	Heated Space Above	4.1
Warehouses 60°F		
All Walls Exposed	Skylights in Roof	5.5
	No Skylights in Roof	5.1
	Heated Space Above	4.0
One Long Warm Common Wall	Skylight in Roof	5.0
	No Skylight in Roof	4.9
	Heated Space Above	3.4
Warm Common Walls on Both Long Sides	Skylight in Roof	4.7
	No Skylight in Roof	4.4
	Heated Space Above	3.0

*Note: This table tends to be conservative particularly for new buildings designed for minimum energy consumption.

Table B - Outside Design Temperature Correction Factor
(for Degrees Fahrenheit)

Outside Design Temp.	50	40	30	20	10	0	-10	-20	-30
Correction Factor	0.29	0.43	0.57	0.72	0.86	1.00	1.14	1.28	1.43

Figure 15.11

Air Conditioning Requirements

BTU's per Hour per S.F. of Floor Area and S.F. per Ton of Air Conditioning

Type Building	BTU per S.F.	S.F. per Ton	Type Building	BTU per S.F.	S.F. per Ton	Type Building	BTU per S.F.	S.F. per Ton
Apartments, Individual	26	450	Dormitory, Rooms	40	300	Libraries	50	240
Corridors	22	550	Corridors	30	400	Low Rise Office, Exterior	38	320
Auditoriums & Theaters	666	18*	Dress Shops	43	280	Interior	33	360
Banks	50	240	Drug Stores	80	150	Medical Centers	28	425
Barber Shops	48	250	Factories	40	300	Motels	28	425
Bars & Taverns	133	90	High Rise Office-Ext. Rms.	46	263	Office (small suite)	43	280
Beauty Parlors	66	180	Interior Rooms	37	325	Post Office, Individual Office	42	285
Bowling Alleys	68	175	Hospitals, Core	43	280	Central Area	46	260
Churches	600	20*	Perimeter	46	260	Residences	20	600
Cocktail Lounges	68	175	Hotel, Guest Rooms	44	275	Restaurants	60	200
Computer Rooms	141	85	Public Spaces	55	220	Schools & Colleges	46	260
Dental Offices	52	230	Corridors	30	400	Shoe Stores	55	220
Dept. Stores, Basement	34	350	Industrial Plants, Offices	38	320	Shop'g. Ctrs., Super Markets	34	350
Main Floor	40	300	General Offices	34	350	Retail Stores	48	250
Upper Floor	30	400	Plant Areas	40	300	Specialty Shops	60	200

*Persons per ton   12,000 BTU = 1 ton of air conditioning

Figure 15.12

Chapter 16
ELECTRICAL

For most interior projects, electrical power will be supplied at least to a panel at or near the space to be constructed. In some cases, for example when renovating existing buildings, a new electric service and feeder distribution system may be required. A new service may include charges to be paid to the local utility, expensive switchgear, and, in some cases, transformers (exterior or within an interior vault). A feeder distribution system involves runs of conduit and large wire to electrical panels located throughout a building (to each floor or tenant space). This type of work should be estimated by an experienced engineer or electrical contractor. Wiring beyond the panel (branch circuits) is usually part of the interior portion of a project. The estimator should be familiar with the various types of distribution systems and methods.

Distribution Systems

For many projects an engineered, specialty distribution network may be in place prior to the start of the interior work. Such networks are often modular and may be installed during construction of the building. In industrial or manufacturing space these networks may be cable tray or bus duct systems (shown in Figure 16.1). These systems are most advantageous where flexibility for power usage and ease of installation are most important. In commercial office buildings, in-place distribution networks may include underfloor raceway systems, trench duct, and cellular concrete floor raceway systems. These are illustrated in Figure 16.2.

Underfloor raceway systems are used extensively in structures that house offices, as they provide an accessible, flexible system for the ever-changing power and relocation needs of an office environment, while remaining visibly unobtrusive. Raceways may be used for electrical power cables, as well as telephone or signal systems wires. They are designed for two types of distribution systems: a two-level system, used when electrical or communications growth is anticipated, and a single-level system, used when such growth is not anticipated and when installation cost is determined by present needs.

Trench duct is a flush electrical raceway which is used as a feeder for cells of a cellular steel floor system or as a distribution duct for an underfloor duct system. Trench duct is also utilized as a self-contained raceway system for computer rooms, research laboratories, hospitals, and other locations where underfloor power and data cables are specified. Trench duct is manufactured in three basic types: assembled bottom type, unassembled bottom type, and unassembled bottomless type.

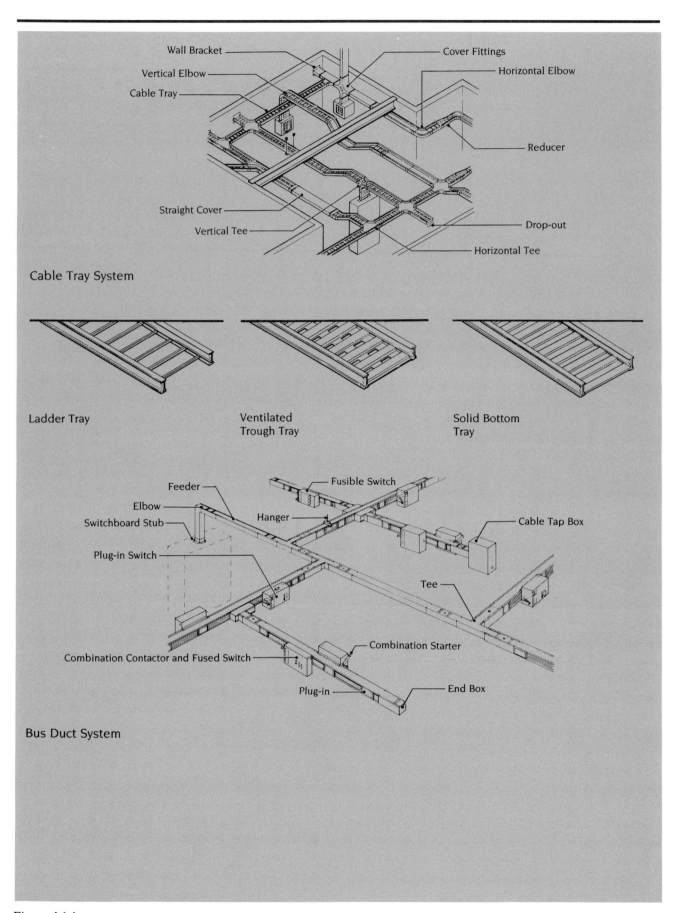

Wall Bracket

Vertical Elbow

Cable Tray

Cover Fittings

Horizontal Elbow

Reducer

Straight Cover

Vertical Tee

Drop-out

Horizontal Tee

Cable Tray System

Ladder Tray

Ventilated
Trough Tray

Solid Bottom
Tray

Feeder

Elbow

Switchboard Stub

Plug-in Switch

Fusible Switch

Hanger

Cable Tap Box

Tee

Combination Contactor and Fused Switch

Combination Starter

Plug-in

End Box

Bus Duct System

Figure 16.1

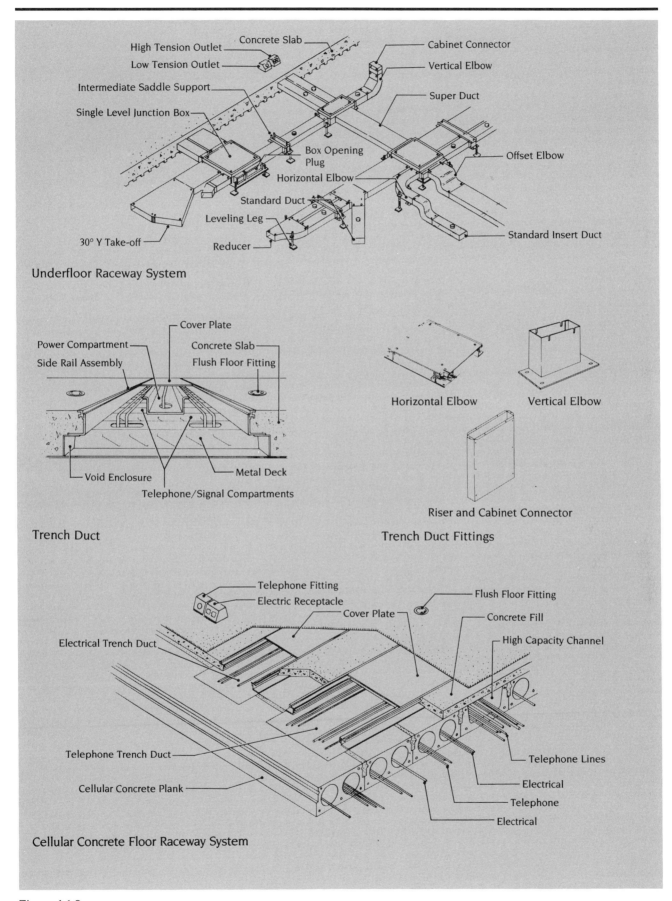

High Tension Outlet — Concrete Slab — Cabinet Connector

Low Tension Outlet — Vertical Elbow

Intermediate Saddle Support — Super Duct

Single Level Junction Box

Box Opening Plug

Horizontal Elbow

Offset Elbow

Standard Duct

Leveling Leg

30° Y Take-off

Reducer

Standard Insert Duct

Underfloor Raceway System

Cover Plate

Power Compartment

Concrete Slab

Side Rail Assembly

Flush Floor Fitting

Void Enclosure

Metal Deck

Telephone/Signal Compartments

Horizontal Elbow   Vertical Elbow

Riser and Cabinet Connector

Trench Duct   **Trench Duct Fittings**

Telephone Fitting

Electric Receptacle

Flush Floor Fitting

Cover Plate

Concrete Fill

Electrical Trench Duct

High Capacity Channel

Telephone Trench Duct

Telephone Lines

Cellular Concrete Plank

Electrical

Telephone

Electrical

Cellular Concrete Floor Raceway System

Figure 16.2

Cellular concrete floor raceways are used to provide a safe and effective underfloor electrical distribution system. The floor structure consists of precast reinforced concrete members that contain smooth, round longitudinal cells which form the raceways for the power and signal wiring systems.

If no "built-in" distribution system is in place within the space to be constructed, certain options exist. Conventional wiring consists primarily of the following types of raceways and/or conductors:

- Rigid galvanized steel conduit
- Aluminum conduit
- EMT (thin wall conduit)
- BX (armored cable)
- Romex (non-metallic sheathed cable)

The use of each type will depend on national and local electrical, building, and fire codes and on occupant requirements. Figure 16.3 demonstrates how each of the different types of wiring can affect the costs.

In recent years, a new type of distribution system has been developed which exhibits the advantages of the underfloor raceway and trench duct systems but allows the flexibility of conventional methods – custom installation when the interior work is performed, and ease of change during operation for specific occupant requirements. This new type is the undercarpet power, data, and communication systems.

Undercarpet systems are an alternative to conventional round cable for wiring commercial and industrial offices. They provide a method of distributing power almost anywhere on the floors without having to channel through underfloor ducts, walls, or ceilings.

The flat, low profile design of this type of system allows for its installation directly on top of wood, concrete, composition, or ceramic floors. It is then covered with carpet squares (18" to 30" square) which allow for change at any time.

The basic elements of undercarpet systems include three groupings of components: specialized flat, low profile cable; transition fittings, which house the round-to-flat conductor connections at the supply end of the cable; and floor fittings, which house the flat-to-round connections and provide various access configurations at the other end of the cable. Undercarpet power, telephone, and data systems are illustrated in Figure 16.4.

These systems involve specialized materials and installation methods. Specific manufacturers should be consulted for material costs. Installation costs (for other than rough budget purposes) should be estimated by experienced subcontractors.

16.8 Special Systems	CREW	DAILY OUTPUT	MAN-HOURS	UNIT	MAT.	LABOR	EQUIP.	TOTAL	TOTAL INCL O&P	
0900 Galvanized steel conduit	1 Elec	6.10	1.310	Ea.	16	30		46	61	810
0980 Exhaust fan wiring										
1000 Using non-metallic sheathed cable	1 Elec	10	.800	Ea.	8	18.10		26.10	35	
1020 BX cable		8.30	.964		12	22		34	45	
1060 EMT conduit		6.70	1.190		15	27		42	56	
1080 Aluminum conduit		5	1.600		18	36		54	72	
1100 Galvanized steel conduit	↓	4.70	1.700	↓	17	39		56	75	
1180 Fire alarm smoke detector & horn										
1200 Using non-metallic sheathed cable	1 Elec	10	.800	Ea.	39	18.10		57.10	69	
1220 BX cable		8.30	.964		43	22		65	79	
1240 EMT conduit		6.70	1.190		44	27		71	88	
1280 Aluminum conduit		5	1.600		51	36		87	110	
1300 Galvanized steel conduit		4.70	1.700		49	39		88	110	
1400 Front doorbell	↓	5.50	1.450	↓	15	33		48	64	
1580 Furnace circuit and switch										
1600 Using non-metallic sheathed cable	1 Elec	6	1.330	Ea.	11	30		41	56	
1620 BX cable		5	1.600		14	36		50	68	
1640 EMT conduit		4	2		18	45		63	85	
1680 Aluminum conduit		3	2.670		23	60		83	115	
1690 Galvanized steel conduit	↓	2.80	2.860	↓	22	65		87	120	
1700 Ground fault receptacle										
1720 Using non-metallic sheathed cable	1 Elec	8	1	Ea.	36	23		59	72	
1740 BX cable		6.60	1.210		41	27		68	85	
1760 EMT conduit		5.40	1.480		44	34		78	97	
1770 Aluminum conduit		4	2		48	45		93	120	
1780 Galvanized steel conduit	↓	3.80	2.110	↓	47	48		95	120	
1790 Heater circuits										
1800 Using non-metallic sheathed cable	1 Elec	8	1	Ea.	8	23		31	42	
1820 BX cable		6.60	1.210		11	27		38	52	
1840 EMT conduit		5.40	1.480		14	34		48	64	
1860 Aluminum conduit		4	2		18	45		63	85	
1880 Galvanized steel conduit	↓	3.80	2.110	↓	17	48		65	88	
1980 Intercom, 8 stations										
2000 Using non-metallic sound cable	1 Elec	.70	11.430	Total	260	260		520	660	
2200 Light fixtures, average	"	16	.500	Ea.	22	11.35		33.35	41	
2380 Lighting wiring										
2400 Using non-metallic sheathed cable	1 Elec	16	.500	Ea.	8	11.35		19.35	25	
2420 BX cable		13.30	.602		12	13.60		25.60	33	
2440 EMT conduit		10.70	.748		15	16.95		31.95	41	
2460 Aluminum conduit		8	1		19	23		42	54	
2480 Galvanized steel conduit	↓	7.50	1.070	↓	18	24		42	55	
2580 Range circuits										
2600 Using non-metallic sheathed cable	1 Elec	4	2	Ea.	30	45		75	99	
2620 BX cable		3.30	2.420		42	55		97	125	
2640 EMT conduit		2.70	2.960		48	67		115	150	
2660 Aluminum conduit		2	4		54	91		145	190	
2680 Galvanized steel conduit		1.90	4.210		51	95		146	195	
2800 Service and panel, 100 amp		1.20	6.670		220	150		370	460	
3000 200 amp	↓	.90	8.890	↓	420	200		620	755	
3180 Switch, single pole										
3200 Using non-metallic sheathed cable	1 Elec	16	.500	Ea.	8	11.35		19.35	25	
3220 BX cable		13.30	.602		12	13.60		25.60	33	
3240 EMT conduit		10.70	.748		14	16.95		30.95	40	
3260 Aluminum conduit		8	1		22	23		45	57	
3280 Galvanized steel conduit	↓	7.50	1.070	↓	19	24		43	56	
3390 Switch, 3-way										
3400 Using non-metallic sheathed cable	1 Elec	12	.667	Ea.	10	15.10		25.10	33	
3420 BX cable		10	.800		15	18.10		33.10	43	
3430 EMT conduit		8	1		17	23		40	52	
3440 Aluminum conduit	↓	6	1.330	↓	24	30		54	70	

For expanded coverage of these items see *Means Electrical Cost Data 1987*

221

Figure 16.3

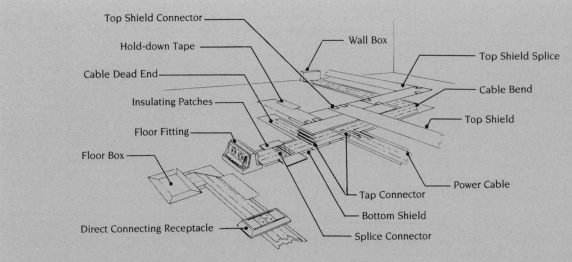

Undercarpet Power Systems

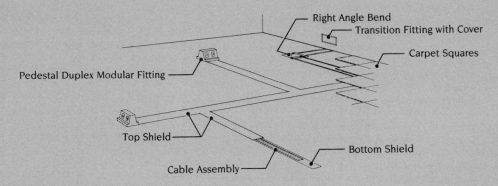

Undercarpet Telephone Systems

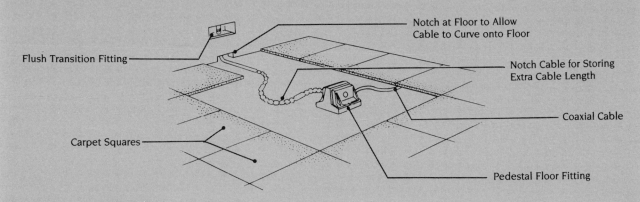

Undercarpet Data System

Figure 16.4

Lighting and Power

There are various types of lighting which may be included in an interior estimate including:

- Fluorescent
- Incandescent
- High Intensity Discharge
- Emergency Lights and Power

Typical fluorescent and incandescent fixtures are illustrated in Figure 16.5. Typical high intensity discharge fixtures are illustrated in Figure 16.6. A graphic relationship of the light output versus power usage for the different types is shown in Figure 16.7.

Fluorescent Lighting

A fluorescent lamp consists of a hot cathode in a phosphor-coated tube which contains inert gas and mercury vapor. When energized, the cathode causes a mercury arc to produce ultraviolet light and fluorescence on the phosphor coating of the tube. The color of the light varies according to the type of phosphor used in the coating. Fluorescent lamps are high in efficiency, and with limited switching on and off, they have a life in excess of 20,000 hours. A ballast is required in the lamp circuit to limit the current. Ballasts are required in various watt-saving types and can be matched with special energy-saving lamps. Special ballasts are required for dimming.

Fluorescent tubes are produced in many different wattages, sizes, and types. One manufacturer lists lamps of 4 watts to 215 watts with lengths of 6" to 96". Three basic types of fluorescent lamps are currently manufactured: preheat, instant start, and rapid start. The preheat lamp, which is the oldest type, requires a starter. The instant start lamp, or slimline, was developed after the preheat type. The rapid start lamp, which is most commonly used today, operates at 425 mA. High output lamps operate at 800 mA; very high output, at 1500 mA. Because the ballasts used in high output and very high output lamps tend to be noisy, these types of fluorescent lamps are not recommended for use in quiet areas.

Incandescent Lighting

An incandescent lamp is a glass bulb which contains a tungsten filament with a mixture of argon and nitrogen gas. The base of the bulb is usually capped with a screw base made of brass or aluminum. Incandescent lamps are versatile sources of light, as they are manufactured in many different sizes, shapes, wattages, and base configurations. Some of these variations include bulbs which feature clear, frosted, and hard glass (for weatherproof applications); aluminized reflectors, wide, narrow, and spot beam pre-focused; and three-way wattage switching. For general applications, incandescent lamps are rated from 2 watts to 1500 watts, but some street lighting lamps may be rated as high as 15,000 watts.

Along with the advantage of variety of lamp sizes and special features, the relatively small size of incandescents allows them to be fit easily into the design of the fixtures which hold them. They are low in cost, soft in color, and easy to use with dimmers. The disadvantages of incandescent lamps include relatively short life, usually less than 1000 hours, and higher energy consumption than fluorescent and mercury vapor lamps (shown in Figure 16.7). If the incandescent lamp is used in a system with a higher voltage than that recommended by the manufacturer, the life of the lamp decreases significantly. An excess of just 10 volts above the recommended voltage can reduce the lamp's life considerably.

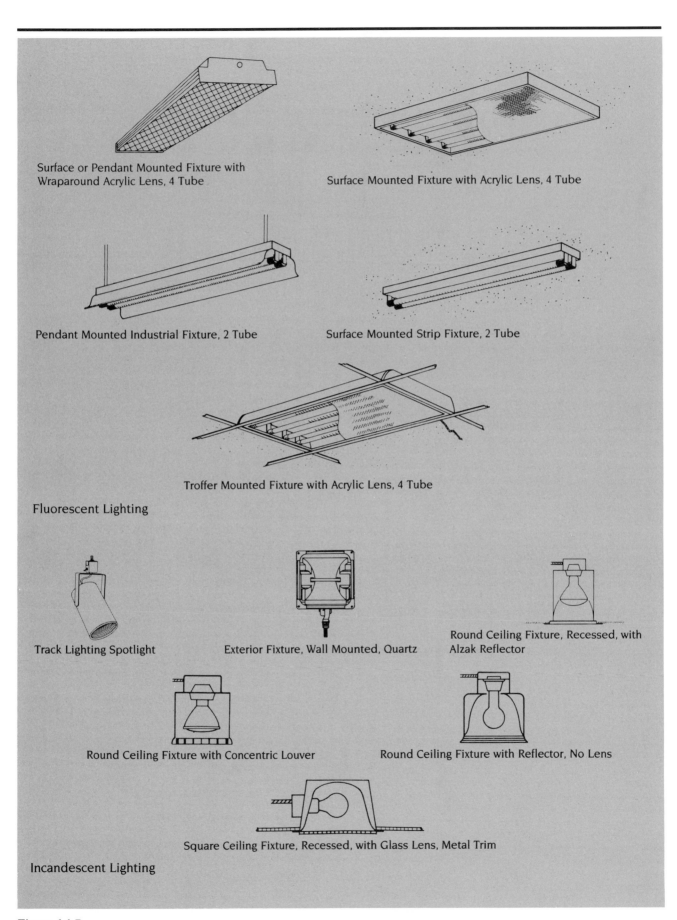

Surface or Pendant Mounted Fixture with Wraparound Acrylic Lens, 4 Tube

Surface Mounted Fixture with Acrylic Lens, 4 Tube

Pendant Mounted Industrial Fixture, 2 Tube

Surface Mounted Strip Fixture, 2 Tube

Troffer Mounted Fixture with Acrylic Lens, 4 Tube

Fluorescent Lighting

Track Lighting Spotlight

Exterior Fixture, Wall Mounted, Quartz

Round Ceiling Fixture, Recessed, with Alzak Reflector

Round Ceiling Fixture with Concentric Louver

Round Ceiling Fixture with Reflector, No Lens

Square Ceiling Fixture, Recessed, with Glass Lens, Metal Trim

Incandescent Lighting

Figure 16.5

160

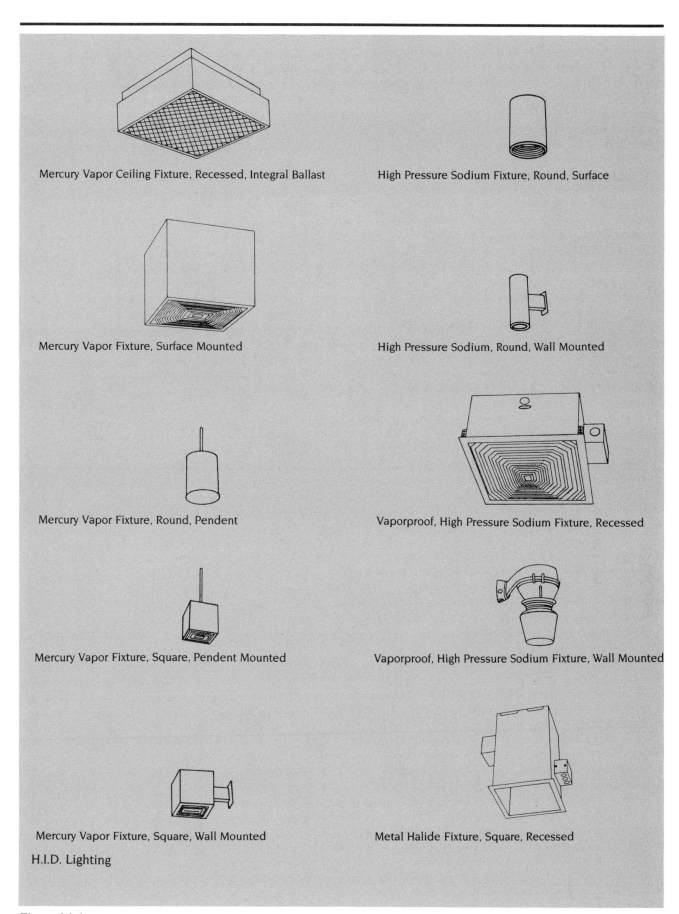

Mercury Vapor Ceiling Fixture, Recessed, Integral Ballast

High Pressure Sodium Fixture, Round, Surface

Mercury Vapor Fixture, Surface Mounted

High Pressure Sodium, Round, Wall Mounted

Mercury Vapor Fixture, Round, Pendent

Vaporproof, High Pressure Sodium Fixture, Recessed

Mercury Vapor Fixture, Square, Pendent Mounted

Vaporproof, High Pressure Sodium Fixture, Wall Mounted

Mercury Vapor Fixture, Square, Wall Mounted

Metal Halide Fixture, Square, Recessed

H.I.D. Lighting

Figure 16.6

The quartz lamp, or tungsten halogen lamp, is a special type of incandescent lamp which consists of a quartz tube with various configurations. Some quartz lamps are simple double-ended tubes, while others include a screw base or are mounted inside an R- or PAR-shaped bulb. A quartz lamp maintains maximum light output throughout its life, which varies according to its type and size from 2000 to 4000 hours. Generally, quartz lamps are more energy efficient than regular incandescents, but their purchase price is higher. Quartz lamps are available in sizes from 36 watts to 1500 watts.

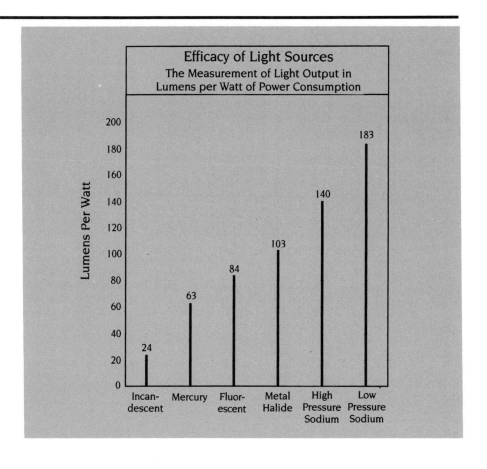

Figure 16.7

Fixtures for regular incandescent lamps vary widely, depending on their function, location, and desired appearance. Simple lampholders, decorative multi-lamp chandeliers, down lights, spotlights, accent wall lights, and track lights are just a few of the many types of fixtures. Some of the fixtures used for quartz lamps include exterior flood, track, accent, and emergency lights, to name a few.

High Intensity Discharge Lighting

High intensity discharge lighting (HID) types include mercury vapor, metal halide, and high and low pressure sodium lamps. HID lamps are usually installed to light large indoor and outdoor areas, such as factories, gymnasiums, sports complexes, parking lots, building perimeters, streets, and highways.

A mercury vapor lamp is a glass bulb which contains high pressure mercury vapor. It works on the same principle as do fluorescent lamps, except that the pressure of the mercury vapor within the bulb is much higher than that of a standard fluorescent tube. A metal halide lamp is basically a mercury vapor lamp with modifications in its arc tube arrangement. Generally, the color, as well as the efficiency, is better in this type of lamp when compared to conventional mercury vapor models.

High pressure sodium lamps differ from mercury lamps primarily in the type of vapor contained in the arc tube. These lamps use a mixture of sodium, mercury, and xenon to produce a slightly yellow color. Like mercury lamps, sodium lamps are efficient in energy consumption and high rated lamp life.

The low pressure sodium lamp is more efficient than the high pressure sodium type, but its intense yellow color makes it unsuitable for indoor use. This type of lamp is used primarily for lighting large outdoor areas, such as roadways, parking lots, and places that require security lighting.

Most projects will include many different types and sizes of lighting fixtures. In order to organize the lighting fixtures on the drawing, fixtures can be given a letter symbol for each type. A lighting schedule is prepared to list each type with a description and specifications. An example of a lighting schedule is shown in Figure 16.8.

There are three basic ways to estimate the costs for lighting and other branch circuit power requirements (receptacles, switches, special circuits). The most accurate is a detailed unit price estimate. This involves counting each fixture or wiring device, measuring the lengths of each type of raceway and conductor, and obtaining material prices and labor productivity rates in order to determine costs. The detail and time required limits this type of estimate to subcontractors, primarily for bidding purposes.

A second method is to use square foot costs that may be developed from past projects or obtained from sources such as Means *Interior Cost Data*, 1987, as shown in Figure 16.9. While this method is fast, it does not allow for variation in fixture or device density unless records are kept for different types of occupancy (e.g., office, school, etc.).

In recent years, most state and local governments have adopted energy codes which, in effect, limit the maximum power usage based on type of occupancy. These limits are usually defined as the maximum wattage per square foot of area. Besides type of building use, different types of spaces within a building may have different requirements: office areas, bathrooms, hallways, etc. Figure 16.10 shows typical allowances based on type of occupancy and use. Local codes and officials should be consulted.

The estimator can use these power usage requirements to the best advantage when determining costs. Device and lighting costs can be determined based on wattage per square foot. This third method of estimating can account for fixture and device density variations based on different types of occupancy and use. Examples of how these costs are developed, for fluorescent lighting and for receptacles are shown in Figures 16.11 and 16.12, respectively.

Type	Manufacturer & Catalog #	Fixture	Type	Lamps Qty	Volts	Watts	Mounting	Remarks
A	Meansco #7054	2′x4′ Troffer	F-40 CW	4	277	40	Recessed	Acrylic Lens
B	Meansco #7055	1′x4′ Troffer	F-40 CW	2	277	40	Recessed	Acrylic Lens
C	Meansco #7709	6″x4′	F-40 CW	1	277	40	Surface	Acrylic Wrap
D	Meansco #7710	6″x8′ Strip	F96T12 CW	1	277	40	Surface	
E	Meansco #7900A	6″x4′	F-40	1	277	40	Surface	Mirror Light
F	Kingston #100A	6″x4′	F-40	1	277	40	Surface	Acrylic Wrap
G	Kingston #110C	′ Strip	F-40 CW	1	277	40	Surface	
H	Kingston #3752	Wallpack	HPS	1	277	150	Bracket	W/Photo Cell
J	Kingston #201-202	Floodlight	HPS	1	277	400	Surface	2′ Below Fascia
K	Kingston #203		HPS	1	277	100	Wall Bracket	
L	Meansco #8100	Exit Light	1-13W 20W T6-1/2	1 / 2	120 / 6½	13 / 20	Surface	
M	Meansco #9000	Battery Unit	Sealed Beam	2	12	18	Wall Mount	12 Volt Unit

Figure 16.8

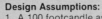

A. Strip Fixture   **B. Surface Mounted**

C. Recessed   **D. Pendent Mounted**

Design Assumptions:
1. A 100 footcandle average maintained level of illumination.
2. Ceiling heights range from 9' to 11'.
3. Average reflectance values are assumed for ceilings, walls and floors.
4. Cool white (CW) fluorescent lamps with 3150 lumens for 40 watt lamps

and 6300 lumens for 8' slimline lamps.
5. Four 40 watt lamps per 4' fixture and two 8' lamps per 8' fixture.
6. Average fixture efficiency values and spacing to mounting height ratios.
7. Installation labor is average U.S. rate as of January 1, 1987.

System Components	QUANTITY	UNIT	COST PER S.F.		
			MAT.	INST.	TOTAL
SYSTEM 09.2-212-0520					
FLUORESCENT FIXTURES MOUNTED 9'-11" ABOVE FLOOR, 100 FC					
TYPE A, 8 FIXTURES PER 400 S.F.					
Steel intermediate conduit, (IMC) ½" diam	.404	L.F.	.24	1.06	1.30
Wire, 600V, type THWN-THHN, copper, solid, #12	.008	C.L.F.	.03	.19	.22
Fluorescent strip fixture 8' long, surface mounted, two 75W SL	.020	Ea.	.88	.84	1.72
Steel outlet box 4" concrete	.020	Ea.	.06	.26	.32
Steel outlet box plate with stud, 4" concrete	.020	Ea.	.02	.07	.09
TOTAL			1.23	2.42	3.65

9.2-212	**Fluorescent Fixtures (by Type)**	COST PER S.F.		
		MAT.	INST.	TOTAL
0520	Fluorescent fixtures, type A, 8 fixtures per 400 S.F.	1.23	2.42	3.65
0560	11 fixtures per 600 S.F.	1.14	2.35	3.49
0600	17 fixtures per 1000 S.F.	1.10	2.31	3.41
0640	23 fixtures per 1600 S.F.	.96	2.20	3.16
0680	28 fixtures per 2000 S.F.	.96	2.20	3.16
0720	41 fixtures per 3000 S.F.	.92	2.19	3.11
0800	53 fixtures per 4000 S.F.	.91	2.14	3.05
0840	64 fixtures per 5000 S.F.	.91	2.14	3.05
0880	Type B, 11 fixtures per 400 S.F.	2.60	3.55	6.15
0920	15 fixtures per 600 S.F.	2.37	3.41	5.78
0960	24 fixtures per 1000 S.F.	2.29	3.40	5.69
1000	35 fixtures per 1600 S.F.	2.13	3.23	5.36
1040	42 fixtures per 2000 S.F.	2.06	3.25	5.31
1080	61 fixtures per 3000 S.F.	2.10	3.12	5.22
1160	80 fixtures per 4000 S.F.	2	3.21	5.21
1200	98 fixtures per 5000 S.F.	1.99	3.18	5.17
1240	Type C, 11 fixtures per 400 S.F.	2.08	3.81	5.89
1280	14 fixtures per 600 S.F.	1.79	3.55	5.34
1320	23 fixtures per 1000 S.F.	1.78	3.53	5.31
1360	34 fixtures per 1600 S.F.	1.69	3.48	5.17
1400	43 fixtures per 2000 S.F.	1.72	3.45	5.17
1440	63 fixtures per 3000 S.F.	1.67	3.40	5.07
1520	81 fixtures per 4000 S.F.	1.61	3.36	4.97
1560	101 fixtures per 5000 S.F.	1.61	3.36	4.97
1600	Type D, 8 fixtures per 400 S.F.	2.09	2.90	4.99
1640	12 fixtures per 600 S.F.	2.09	2.89	4.98
1680	19 fixtures per 1000 S.F.	1.99	2.81	4.80
1720	27 fixtures per 1600 S.F.	1.84	2.75	4.59
1760	34 fixtures per 2000 S.F.	1.83	2.72	4.55
1800	48 fixtures per 3000 S.F.	1.73	2.64	4.37
1880	64 fixtures per 4000 S.F.	1.73	2.64	4.37
1920	79 fixtures per 5000 S.F.	1.73	2.64	4.37

346

Figure 16.9

General Lighting Loads by Occupancies

Type of Occupancy	Unit Load per S.F. (Watts)
Armories and Auditoriums	1
Banks	5
Barber Shops and Beauty Parlors	3
Churches	1
Clubs	2
Court Rooms	2
*Dwelling Units	3
Garages — Commercial (storage)	1/2
Hospitals	2
*Hotels and Motels, including apartment houses without provisions for cooking by tenents	2
Industrial Commercial (loft) Buildings	2
Lodge Rooms	1½
Office Buildings	5
Restaurants	2
Schools	3
Stores	3
Warehouses (storage)	1/4
*In any of the above occupancies except one-family dwellings and individual dwelling units of multi-family dwellings:	
Assembly Halls and Auditoriums	1
Halls, Corridors, Closets	1/2
Storage Spaces	1/4

Lighting Limit (Connected Load) for Listed Occupancies: New Building Proposed Energy Conservation Guideline

Type of Use	Maximum Walls per S.F.
Interior:	
Category A: Classrooms, office areas, automotive mechanical areas, museums, conference rooms, drafting rooms, clerical areas, laboratories, merchandising areas, kitchens, examining rooms, book stacks, athletic facilities.	3.00
Category B: Auditoriums, waiting areas, spectator areas, restrooms, dining areas, transportation terminals, working corridors in prisons and hospitals, book storage areas, active inventory storage, hospital bedrooms, hotel and motel bedrooms, enclosed shopping mall concourse areas, stairways.	1.00
Category C: Corridors, lobbies, elevators, inactive storage areas.	0.50
Category D: Indoor parking.	0.25
Exterior:	
Category E: Building perimeter: wall-wash, facade, canopy.	5.00 (per linear foot)
Category F: Outdoor parking.	0.10

Figure 16.10

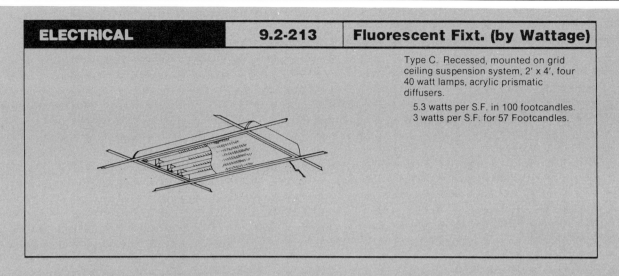

Type C. Recessed, mounted on grid ceiling suspension system, 2' x 4', four 40 watt lamps, acrylic prismatic diffusers.

5.3 watts per S.F. in 100 footcandles.
3 watts per S.F. for 57 Footcandles.

System Components	QUANTITY	UNIT	COST PER S.F.		
			MAT.	INST.	TOTAL
SYSTEM 09.2-213-0200					
FLUORESCENT FIXTURES RECESS MOUNTED IN CEILING					
1 WATT PER S.F., 20 FC, 5 FIXTURES PER 1000 S.F.					
Steel intermediate conduit, (IMC) ½" diam	.128	L.F.	.08	.34	.42
Wire, 600 volt, type THW, copper, solid, #12	.003	C.L.F.	.01	.07	.08
Fluorescent fixture, recessed, 2'x4', four 40W, w/lens, for grid ceiling	.005	Ea.	.29	.29	.58
Steel outlet box 4" square	.005	Ea.	.01	.07	.08
Fixture whip, Greenfield w/#12 THHN wire	.005	Ea.		.02	.02
TOTAL			.39	.79	1.18

9.2-213	Fluorescent Fixtures (by Wattage)	COST PER S.F.		
		MAT.	INST.	TOTAL
0190	Fluorescent fixtures recess mounted in ceiling			
0200	1 watt per S.F., 20 FC, 5 fixtures per 1000 S.F.	.39	.79	1.18
0240	2 watts per S.F., 40 FC, 10 fixtures per 1000 S.F.	.78	1.53	2.31
0280	3 watts per S.F., 60 FC, 15 fixtures per 1000 S.F	1.17	2.32	3.49
0320	4 watts per S.F., 80 FC, 20 fixtures per 1000 S.F.	1.56	3.08	4.64
0400	5 watts per S.F., 100 FC, 25 fixtures per 1000 S.F.	1.95	3.86	5.81

347

Figure 16.11

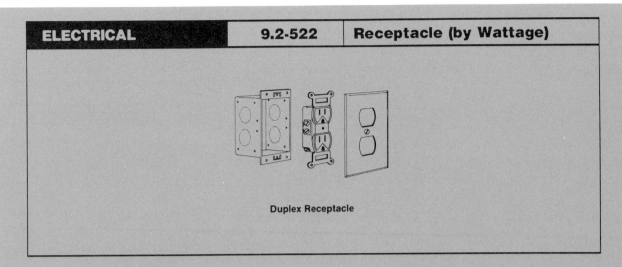

Duplex Receptacle

System Components	QUANTITY	UNIT	COST PER S.F.		
			MAT.	INST.	TOTAL
SYSTEM 09.2-522-0200					
RECEPTACLES INCL. PLATE, BOX, CONDUIT, WIRE & TRANS. WHEN REQUIRED					
2.5 PER 1000 S.F., .3 WATTS PER S.F.					
Steel intermediate conduit, (IMC) ½″ diam	167.000	L.F.	.10	.44	.54
Wire 600V type THWN-THHN, copper solid #12	3.382	C.L.F.	.01	.08	.09
Wiring device, receptacle, duplex, 120V grounded, 15 amp	2.500	Ea.	.01	.02	.03
Wall plate, 1 gang, brown plastic	2.500	Ea.		.01	.01
Steel outlet box 4″ square	2.500	Ea.		.03	.03
Steel outlet box 4″ plaster rings	2.500	Ea.		.01	.01
TOTAL			.13	.59	.72

9.2-522	Receptacle (by Wattage)	COST PER S.F.		
		MAT.	INST.	TOTAL
0190	Receptacles include plate, box, conduit, wire & transformer when required			
0200	2.5 per 1000 S.F., .3 watts per S.F.	.13	.59	.72
0240	With transformer	.15	.62	.77
0280	4 per 1000 S.F., .5 watts per S.F.	.15	.69	.84
0320	With transformer	.19	.73	.92
0360	5 per 1000 S.F., .6 watts per S.F.	.18	.81	.99
0400	With transformer	.22	.87	1.09
0440	8 per 1000 S.F., .9 watts per S.F.	.20	.90	1.10
0480	With transformer	.26	.98	1.24
0520	10 per 1000 S.F., 1.2 watts per S.F.	.22	.97	1.19
0560	With transformer	.32	1.10	1.42
0600	16.5 per 1000 S.F., 2.0 watts per S.F.	.28	1.22	1.50
0640	With transformer	.46	1.44	1.90
0680	20 per 1000 S.F., 2.4 watts per S.F.	.31	1.32	1.63
0720	With transformer	.51	1.58	2.09

350

Figure 16.12

Part II
THE
ESTIMATING
PROCESS

Chapter 17

BEFORE STARTING THE ESTIMATE

The estimating prices and practices in this book are presented primarily from the interior contractor's point of view, however, the information is important to all parties involved in interior construction. For an owner or designer to properly determine budget costs, they must understand how a contractor approaches a project and performs an estimate. Only with mutual understanding of the other's roles can each be sure that estimates are as complete and accurate as possible. This chapter is written from the contractor's point of view and assumes that the estimate is for bidding purposes and that complete plans and specifications are available.

The first step before starting the estimate is to obtain copies of the plans and specifications in *sufficient quantities*. Most estimators mark up plans with colored pencils and make numerous notes and references. For one estimator to work from plans that have been used by another is difficult at best and may easily lead to errors. When more than one estimator works on a particular project, especially if they are working from the same plans, careful coordination is required to prevent omissions as well as duplications. Neatness and color coding of the various items can be used to mark the takeoff progress and to allow a second estimator to finish a takeoff.

The estimator should be aware of and note any instructions to bidders, which may be included in the specifications or in a separate document. To avoid future confusion, the bid's due date, time and place should be clearly stated and understood upon receipt of the construction documents (plans and specifications). The due date should be marked on a calendar and a schedule. Completion of the takeoff should be made as soon as possible, not two days prior to the bid deadline.

If bid security, or a bid bond, is required, then time must be allowed for arrangements to be made, especially if the bonding capability or capacity of a contractor is limited or has not previously been established. If a bond or bid security is to be based on a percentage of the bid, then a preliminary estimate must be prepared to figure the amount of the bid security.

The estimator should attend all pre-bid meetings with the owner or architect, preferably *after* reviewing the plans and specifications. Important points are often brought up at such meetings and details clarified. Attendance is important, not only to show interest in the project, but also to assure equal and competitive bidding. For many projects, attendance is required or the bid will not be accepted. It is to the estimator's advantage to examine and review the plans and specifications before any such meetings and before the initial site visit. It is important to become familiar with the project as soon as possible.

All contract documents should be read thoroughly. They exist to protect all parties involved in the construction process. If not provided, the estimator should make a list of all appropriate documents to be checked with the designer and listed in the bid. The contract documents are written so that the contractors will be bidding equally and competitively, and to ensure that all items in a project are included. The contract documents protect the designer (the architect and/or engineer) by ensuring that all work is supplied and installed as specified. The owner also benefits from thorough and complete construction documents by being assured of a quality job and a complete, functional project. Finally, the contractor benefits from good contract documents because the scope of work is well defined, eliminating the gray areas of what is implied but not stated. "Extras" are more readily avoided. Change orders, if required, are accepted with less argument if the original contract documents are complete, well stated, and most importantly, read by all concerned parties.

During the first review of the specifications, any items to be priced should be identified and noted. All work to be sub-contracted should be noted. "Work by Others" or "Not in Contract" should be clearly defined on the drawings. Certain materials are sometimes specified by the designer and purchased by the owner to be installed (labor only) by the contractor. These items should be noted and the responsibilities of each party clearly understood. An example of such an item might involve allocating responsibility for receiving, temporary storage, and maintenance of restaurant booths furnished by a supplier but installed by the contractor.

The "General Conditions", "Supplemental Conditions" and "Special Conditions" sections of the specifications should be examined carefully by *all* parties involved in the project. These sections describe the items that have a direct bearing on the proposed project, but may not be part of the actual, physical installation. Temporary power and lighting are examples of these kinds of items. Also included in these sections is information regarding completion dates, payment schedules (e.g. retainage), submittal requirements, allowances, alternates, and other important project and contractual requirements. Each of these conditions can have a significant bearing on the ultimate cost of the project. They must be read and understood prior to performing the estimate. Standardized General Conditions are used widely throughout the industry. The supplemental or special conditions, however, are usually prepared for a specific job and must be studied carefully.

While analyzing the plans and specifications, the estimator should evaluate the different portions of the project to determine which areas warrant the most attention. The estimator should focus first on those items which represent the largest cost centers of the project (possible finishes, or millwork) or which entail the greatest risk. These cost centers are not always the portions of the job that require the most time to estimate but are those items that will have the most significant impact on the estimate.

When the overall scope of the work has been identified, the drawings should be examined to confirm the information in the specifications. This is the time to clarify details while reviewing the general content. The estimator should note which sections, elevations and detail drawings are for which plans. At this point and throughout the whole estimating process, the estimator should note and list any discrepancies between the plans and specifications, as well as any possible omissions. It is often stated in bid documents that bidders are obliged to notify the owner or designer of any such discrepancies. When so notified, the designer will most often issue an addendum to the contract documents in order to properly notify all parties concerned and to assure equal and competitive bidding. Competition can be fair only if the same information is provided for all bidders.

Once familiar with the contract documents, the estimator should solicit bids by notifying appropriate subcontractors, manufacturers, and vendors. Those whose work may be affected by the existing building conditions should accompany the estimator on a job site visit (especially in cases of renovation and remodeling, where existing conditions can have a significant effect on the cost of a project).

During a site visit, the estimator should take notes, and possibly photographs, of all situations pertinent to the construction and, thus, to the project estimate. If unusual site conditions exist, or if questions arise during the quantity takeoff, a second site visit is recommended.

In some areas, questions are likely to arise that cannot be answered clearly by the plans and specifications. It is crucial that the owner or responsible party be notified quickly, preferably in writing, so that these questions may be resolved before unnecessary problems arise. Often such items involve more than one contractor and can only be resolved by the owner or designer. A proper estimate cannot be completed until all such questions are answered.

Chapter 18
THE QUANTITY TAKEOFF

The quantity takeoff – the counting of units – is the basis of estimating. To effectively cover all aspects of the project and to include all items, certain steps should be followed. The quantity takeoff should be organized so that the information gathered can be used to future advantage. Scheduling can be made easier if items are taken off and listed by construction phase, or by floor. Material purchasing will similarly benefit. Units for each item should be used consistently throughout the whole project – from takeoff to cost control. In this way, the original estimate can be equitably compared to any progress reports and final cost reports. These methods make keeping track of a job easier.

Part I of this book is devoted to descriptions of the components of interior construction. In that section, takeoff procedures are suggested for the various components. Typical units are also given for each component installation. Items that are generally included with the component are discussed. Also given are the units by which the component is measured or counted.

When working with the plans during the quantity takeoff, consistency is a very important consideration. If each job is approached in the same manner, a pattern will develop, such as moving from the lower floors to the top, clockwise or counterclockwise. The choice of method is not important, but consistency is. The purpose of being consistent is to avoid duplications as well as omissions and errors. Pre-printed forms provide an excellent means for developing consistent patterns. Figures 18.1 to 18.3 are examples of such forms. The Quantity Sheet (Figure 18.1) is designed purely for quantity accumulation. Note that one list of materials and dimensions can be used for up to four different areas or segments of work. Figure 18.2 shows a Cost Analysis sheet, which can be used in conjunction with a quantity sheet. Totals of quantities are transferred to this sheet for pricing and extensions. Figure 18.3, a Consolidated Estimate sheet, is designed to be used for both quantity takeoff and pricing on one form. There are many other variations. Part Three of this book (the sample estimates) contains examples of the use of these forms.

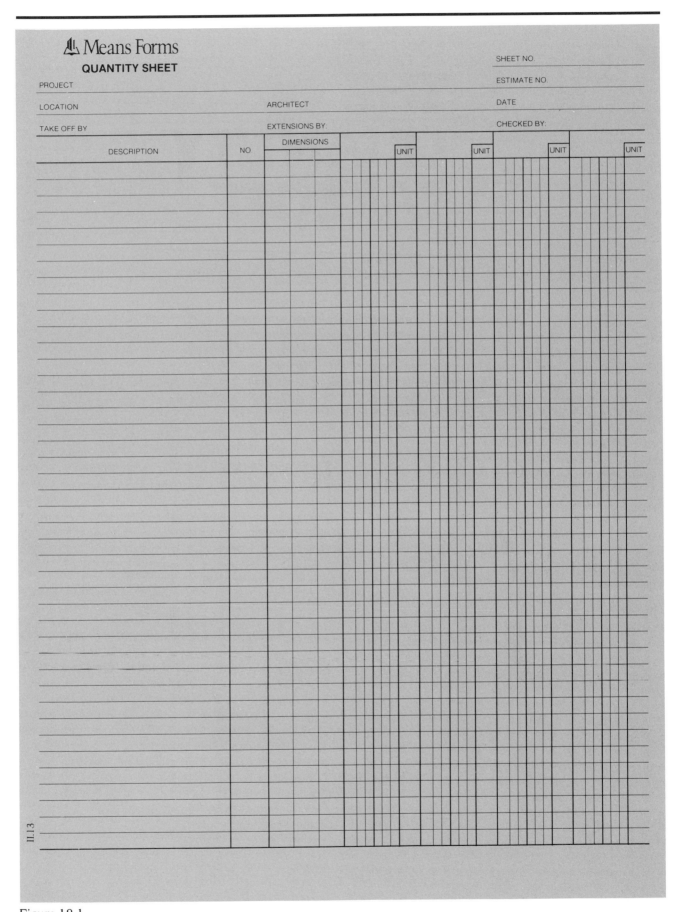

Figure 18.1

176

Means Forms

COST ANALYSIS

SHEET NO.

PROJECT _____ ESTIMATE NO. _____

ARCHITECT _____ DATE _____

TAKE OFF BY: _____ QUANTITIES BY: _____ PRICES BY: _____ EXTENSIONS BY: _____ CHECKED BY: _____

DESCRIPTION	SOURCE/DIMENSIONS			QUANTITY	UNIT	MATERIAL		LABOR	
						UNIT COST	TOTAL	UNIT COST	TOTAL

IL71

Figure 18.2

177

Means Forms

CONSOLIDATED ESTIMATE

SHEET NO.

PROJECT _____ ESTIMATE NO.

ARCHITECT _____ DATE

TAKE OFF BY | QUANTITIES BY | PRICES BY | EXTENSIONS BY | CHECKED BY

DESCRIPTION	NO.	DIMENSIONS			QUANTITIES		UNIT			MATERIAL		LABOR	
						UNIT			UNIT	UNIT COST	TOTAL	UNIT COST	TOTAL

Il.53

Figure 18.3

178

Every contractor might benefit from designing custom company forms. If employees of a company all use the same types of forms, then communications and coordination of the estimating process will proceed more smoothly. One estimator will be able to more easily understand the work of another. R.S. Means has published a book completely devoted to forms and their use, entitled *Means Forms for Building Construction Professionals*. Scores of forms, examples, and instructions for use are included.

Perhaps the two most important forms the interior estimator can use are the Door and Frame Schedule and the Room Finish Schedule (shown in Figures 18.4 and 18.5, respectively). These schedules help to efficiently organize much data involved in a project, for the following reasons:

- The designer can be sure that all work is included in the drawings and specifications.
- The estimator can be sure that all items are included in the estimate.
- The contractor can be sure of proper installation instructions.

Appropriate and easy-to-use forms are the first, and most important, of the "tools of the trade" for estimators. Other tools useful to the estimator may include scales, rotometers, mechanical counters, and colored pencils.

A number of short cuts can be used for the quantity takeoff. If approached logically and systematically, these techniques help to save time without sacrificing accuracy. Consistent use of accepted abbreviations saves the time of writing things out. An abbreviations list might be posted in a conspicuous place to provide a consistent pattern of definitions for use within an office.

All dimensions — whether printed, measured, or calculated — that can be used for determining quantities of more than one item should be listed on a separate sheet and posted for easy reference. Posted gross dimensions can also be used to quickly check for order of magnitude errors.

Rounding off, or decreasing the number of significant digits, should be done only when it will not statistically affect the resulting product. The estimator must use good judgment to determine instances when rounding is appropriate. An overall two or three percent variation in a competitive market can often be the difference between winning a contract or losing a job, or between profit or no profit. The estimator should establish rules for rounding to achieve a consistent level of precision. In general, it is best not to round numbers until the final summary of quantities.

The final summary is also the time to convert units of measure into standards for practical use (square feet of wallcovering to number of rolls, for example). This is done to keep the quantities equitable to what will be purchased and handled.

Be sure to quantify (count) and include "labor only" items that are not shown on the plans. Such items may or may not be indicated in the specifications and might include cleanup, special labor for handling materials, furniture set-up, etc.

Means Forms

**DOOR
AND FRAME SCHEDULE**

PROJECT _____ ARCHITECT _____

LOCATION _____ OWNER _____

PAGE _____ OF _____

DATE _____

BY _____

DOOR NO.	DOOR							FRAME					FIRE RATING		HARDWARE		REMARKS
	SIZE			MAT.	TYPE	GLASS	LOUVER	MAT.	TYPE	DETAILS			LAB	CON	SET NO	KEYSIDE ROOM NO	
	W	H	T							JAMB	HEAD	SILL					

Figure 18.4

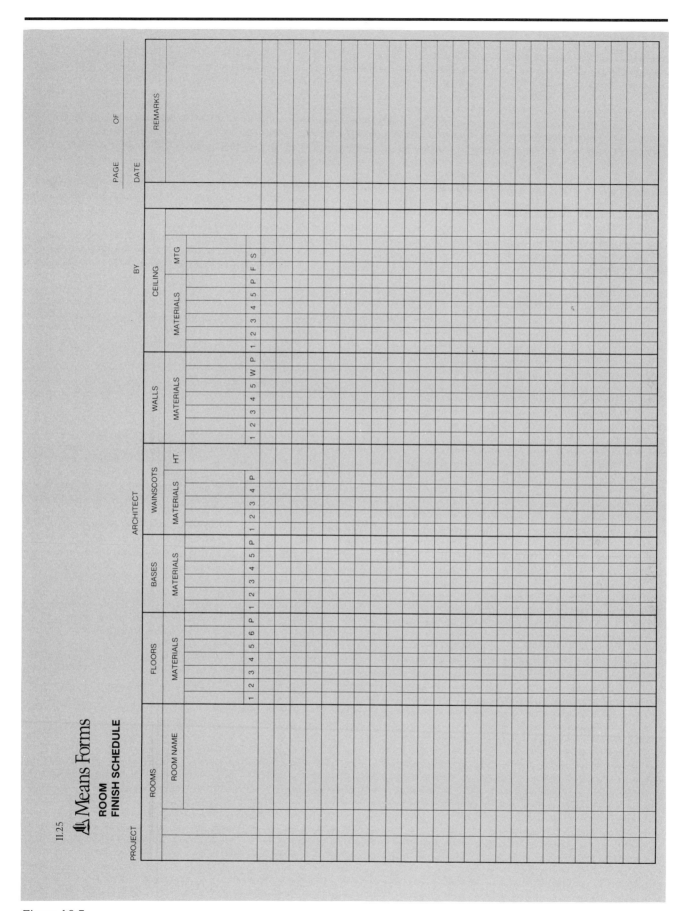

Figure 18.5

The following list is a summation of the suggestions previously mentioned plus a few more guidelines which will be helpful during the quantity takeoff:

- Use preprinted forms.
- Transfer carefully when copying numbers from one sheet to the next.
- List dimensions (width, length) in a consistent order.
- Verify the scale of drawings before using them as a basis for measurement.
- Mark drawings neatly and consistently as quantities are counted.
- Be alert for changes in scale, or notes such as "N.T.S." (not to scale). Sometimes drawings have been photographically reduced.
- Include required items which may not appear in the plans and specs.

The four most important points:

- Print legibly.
- Be organized.
- Use common sense.
- Be consistent.

Chapter 19

PRICING THE ESTIMATE

When the quantities have been counted, values, in the form of unit costs, must be applied and mark-ups (e.g., overhead and profit) added in order to determine the total selling price (the quote). Depending upon the chosen estimating method and the level of detail required, these unit costs may be direct or "bare", or may include overhead, profit or contingencies. In unit price estimating, the unit costs most commonly used are "bare", or "unburdened". Items such as overhead and profit are usually added to the total bare costs at the estimate summary. Discussion of the different types of estimates is included in the Introduction to this book.

Sources of Cost Information

One of the most difficult aspects of the estimator's job is determining accurate and reliable bare cost data. Sources for such data are varied, but can be categorized in terms of their relative reliability. The most reliable of any cost information is direct quotations from usual sources of supply (dealers, distributors, factory reps) for a particular job, and accurate, up-to-date, well-kept records of completed work by the estimator's own company. There is no better cost for a particular construction item than the *actual* cost to the contractor of that item from another recent job, modified (if necessary) to meet the requirements of the project being estimated.

Bids from responsible subcontractors are the next most reliable source of cost data. Any estimating inaccuracies are essentially absorbed by the subcontractor. A subcontract bid is a known, fixed cost, prior to the project. Whether the price is "right" or "wrong" does not matter (as long as it is a responsible competitive bid with no apparent errors). The bid is what the appropriate portion of work will cost. No estimating is required, except for possible verification of the quote and comparison with other subcontractors' quotes.

Quotations by vendors for material costs are, for the same reasons, as reliable as subcontract bids. In this case, however, the estimator must apply estimated labor costs. Thus the "installed" price for a particular item may be more variable.

Whenever possible, all price quotations from vendors or subcontractors should be confirmed in writing. Qualifications and exclusions should be clearly stated. The items quoted should be checked to be sure that they are complete and as specified. One way to assure these requirements is to prepare a form on which all subcontractors and vendors must submit their quotations. This form, generally called a "request for quote", can suggest all of the appropriate questions. It also provides a format to organize the information needed by the estimator.

The above procedures are ideal, but in the realistic haste of estimating and bidding, quotations are often received verbally, either in person or by telephone. The importance of gathering all pertinent information is heightened because omissions are more likely. A preprinted form, such as the one shown in Figure 19.1, is essential to assure that all required information and qualifications are obtained and understood. How often has the subcontractor or vendor said, "I didn't know that I was supposed to include that"? With the help of such forms, the appropriate questions will be asked.

If the estimator has no cost records for a particular item and is unable to obtain a quotation, then another reliable source of price information is the use of current construction cost books such as Means *Interior Cost Data*. R. S. Means presents all such data in the form of national averages; these figures can be adjusted to local conditions, a procedure that will be explained later in this book. In addition to being a source of primary costs, current annual construction cost data books can also be useful as reference tools or as cross-checks for verifying costs obtained from unfamiliar sources.

Lacking cost information from any of the above-mentioned sources, the estimator may have to rely on experience and personal knowledge of the field to develop costs.

No matter which source of cost information is used, the system and sequence of pricing should be the same as that used for the quantity takeoff. This consistent approach should continue through both accounting and cost control during work on the project.

Types of Costs

All costs included in a unit price estimate can be divided into two types: direct and indirect. Direct costs are those dedicated solely to the physical construction of a specific project. Material, labor, equipment, and subcontract costs, as well as project overhead costs, are all direct.

Indirect costs are usually added to the estimate at the summary stage and are most often calculated as percentages of the direct costs. They include such items as taxes, insurance, overhead, profit, and contingencies. The indirect costs account for great variation in estimates among different bidders.

Types of Costs in a Construction Estimate

Direct Costs	Indirect Costs
Material	Taxes
Labor	Insurance
Equipment	Office Overhead
Subcontractors	Profit
Project Overhead	Contingencies
Sales tax	
Bonds	

A clear understanding of direct and indirect cost factors is a fundamental part of pricing the estimate. The following text addresses the components of direct and indirect costs in detail.

⚓ Means Forms

**TELEPHONE
QUOTATION**

	DATE
PROJECT	TIME
FIRM QUOTING	PHONE ( )
ADDRESS	BY
ITEM QUOTED	RECEIVED BY

WORK INCLUDED	AMOUNT OF QUOTATION

DELIVERY TIME		**TOTAL BID**	
DOES QUOTATION INCLUDE THE FOLLOWING:		If ☐ NO is checked, determine the following:	
STATE & LOCAL SALES TAXES	☐ YES ☐ NO	MATERIAL VALUE	
DELIVERY TO THE JOB SITE	☐ YES ☐ NO	WEIGHT	
COMPLETE INSTALLATION	☐ YES ☐ NO	QUANTITY	
COMPLETE SECTION AS PER PLANS & SPECIFICATIONS	☐ YES ☐ NO	DESCRIBE BELOW	
EXCLUSIONS AND QUALIFICATIONS			

ADDENDA ACKNOWLEDGEMENT	**TOTAL ADJUSTMENTS**	
	ADJUSTED TOTAL BID	

II.153

Figure 19.1

185

Chapter 20
DIRECT COSTS

Direct costs can be defined as those necessary for the completion of the project, in other words, the hard costs. Material, labor, and equipment are among the more obvious items in this category. While subcontract costs include the overhead and profit (indirect costs) of the subcontractor, they are considered to be direct costs to the prime contractor. Also included are certain project overhead costs for items that may be necessary for construction. Examples are rolling scaffolding, tools, and temporary power and lighting. Sales tax and bonds are additional direct costs, since they are essential for the performance of the project.

Material

When quantities have been carefully taken off, estimates of material cost can be very accurate. For a high level of accuracy, the material unit prices (especially for expensive finish materials) must be reliable and current. The most reliable source of material costs is a quotation from a vendor for the particular job in question. Ideally, the vendor should have access to the plans and specifications for verification of quantities and specified products. Providing material quotes (and specified submittals for approval) is a service which most suppliers will readily perform.

Material pricing appears relatively simple and straightforward. There are, however, certain considerations that the estimator must address when analyzing material quotations. The reputation of the vendor is a significant factor. Can the vendor "deliver", both figuratively and literally? Estimators may choose not to rely on a "competitive" lower price from an unknown vendor, but will instead use a slightly higher price from a known, reliable vendor. Experience is the best judge for such decisions.

There are many other questions that the estimator should ask. How long is the price guaranteed? When does the price guarantee begin? Is there an escalation clause? Does the price include delivery charges or sales tax, if required? Where is the point of FOB? This can be an extremely important factor. Are product guarantees and warranties in compliance with the specification requirements? Will there be adequate and appropriate storage space available? If not, can staggered shipments be made? Note that most of these questions can be addressed on the form in Chapter 19, Figure 19.1. More information should be obtained, however, to assure that a quoted price is accurate and competitive.

The estimator must be sure that the quotation or obtained price is for the materials according to the plans and specifications. Architects, engineers, and designers may write into the specifications that:

- A particular type or brand of product must be used with no substitution.
- A particular type or brand of product is specified, but alternate brands of equal quality and performance may be accepted *upon approval.*
- No particular type or brand is specified.

Depending on the options, the estimator may be able to find an acceptable, less expensive alternative. In some cases, these substitutions can substantially lower the cost of a project. Note also that many specification packages will require that "catalogue cuts" be submitted for approval for certain materials as part of the bid proposal. In this case, there is pressure on the estimator to obtain the lowest possible price on materials that will meet the specified criteria.

When the estimator has received material quotations, there are still other considerations which should have a bearing on the final choice of a vendor. Lead time – the amount of time between order and delivery – must be determined and considered. It does not matter how competitive or low a quote is if the material cannot be delivered to the job site on time to support the schedule. If a delivery date is promised, is there a guarantee, or a penalty clause for late delivery?

The estimator should also determine if there are any unusual payment requirements. Cash flow for a company can be severely affected if a large material purchase, thought to be payable in 30 days, is delivered C.O.D. Truck drivers may not allow unloading until payment has been received. Such requirements for possible financing must be determined during the estimating stage so that the cost of borrowing money, if necessary, can be included.

If unable to obtain the quotation of a vendor from whom the material would be purchased, the estimator has other sources for obtaining material prices. These include, in order of reliability:

1. *Current* price lists from manufacturers' catalogs. Be sure to check that the list is for "contractor prices".
2. Cost records from previous jobs. Historical costs must be updated for present market conditions.
3. Reputable and current annual unit price cost books, such as Means *Interior Cost Data*. Such books usually represent national averages and must be factored to local markets.

No matter which price source is used, the estimator must be sure to include an allowance for any burdens, such as delivery costs, taxes or finance charges, over the actual cost of the material. These same concerns mentioned previously that apply to vendor quotations should be taken into consideration when using these other price sources.

Labor

In order to determine the installation cost for each item of construction, the estimator must know two pieces of information: first, the labor rate (hourly wage or salary) of the worker, and second, how much time a worker will need to complete a given unit of the installation – in other words, the productivity or output. Wage rates are usually known going into a project, but productivity may be very difficult to determine.

To estimators working for contractors, the construction labor rates that the contractor pays will be known, well-documented and constantly updated. Projected labor rates for construction jobs of long duration should also be contended with. Estimators for owners, architects, or engineers must determine labor rates from outside sources. Unit price data books, such as Means *Interior Cost Data* provide national average labor wage rates. The unit costs for labor are based on these averages. Figure 20.1 shows national average *union* rates for the construction industry based on January 1, 1987. Figure 20.2 lists national average *non-union* rates, again based on January 1, 1987.

If more accurate union labor rates are required, the estimator has different options. Union locals can provide rates (as well as negotiated increases) for a particular location. Employer bargaining groups can usually provide labor cost data as well. R. S. Means Co., Inc. publishes *Labor Rates for the Construction Industry* on an annual basis. This book lists the union labor rates by trade for over 300 U.S. and Canadian cities.

Determination of non-union, or "open shop" rates is much more difficult. In larger cities, employer organizations often exist to represent non-union contractors. These organizations may have records of local pay scales, but ultimately, wage rates are determined by individual contractors.

Productivity is the least predictable of all factors for interior construction. It is important to determine as accurately as possible prevailing productivity. The best source of labor productivity or labor units (and therefore labor costs) is the estimator's well-kept records from previous projects. If there are no company records for productivity, cost data books, such as Means *Interior Cost Data* and productivity reference books, such as *Means Man-Hour Standards* can be invaluable. Included with the listing for each individual construction item is the designation of a suggested crew make-up. The crew is the minimum grouping of workers which can be expected to accomplish the task efficiently. Figure 20.3, a typical page from Means *Interior Cost Data* includes this data and indicates the manhours as well as the average daily output of the designated crew for each construction item.

The estimator who has neither company records nor the sources described above must put together the appropriate crews and determine the expected output or productivity. This type of estimating should only be attempted based upon strong experience a id considerable exposure to construction methods and practices. There are rare occasions when this approach is necessary to estimate a particular item or a new technique. Even then, the new labor units are often extrapolated from existing figures for similar work, rather than being created from scratch.

Installing Contractor's Overhead & Profit

Below are the **average** installing contractor's percentage mark-ups applied to base labor rates to arrive at typical billing rates.

Column A: Labor rates are based on union wages averaged for 30 major U.S. cities. Base rates including fringe benefits are listed hourly and daily. These figures are the sum of the wage rate, employer-paid fringe benefits such as vacation pay, employer-paid health and welfare costs, pension costs, plus appropriate training and industry advancement funds costs.

Column B: Workers' Compensation rates are the national average of state rates established for each trade.

Column C: Column C lists average fixed overhead figures for all trades. Included are Federal and State Unemployment costs set at 5.5%; Social Security Taxes (FICA) set at 7.15%; Builder's Risk Insurance costs set at 1.14%; and Public Liability costs set at 0.82%. All the percentages except those for Social Security Taxes vary from state to state as well as from company to company.

Column D and E: Percentages in Columns D and E are based on the presumption that the installing contractor has annual billing of $500,000 and up. Overhead percentages may increase with smaller annual billing. The overhead percentages for any given contractor may vary greatly and depend on a number of factors, such as the contractor's annual volume, engineering and logistical support costs, and staff requirements. The figures for overhead and profit will also vary depending on the type of job, the job location, and the prevailing economic conditions. All factors should be examined very carefully for each job.

Column F: Column F lists the total of columns B, C, D, and E.

Column G: Column G is Column A (hourly base labor rate) multiplied by the percentage in Column F (O&P percentage).

Column H: Column H is the total of Column A (hourly base labor rate) plus Column G (Total O&P).

Column I: Column I is Column H multiplied by eight hours.

		A		B	C	D	E	F		G	H	I
		Base Rate Incl. Fringes		Workers' Comp. Ins.	Average Fixed Overhead	Over-head	Profit	Total Overhead & Profit			Rate with O & P	
Abbr.	Trade	Hourly	Daily					%	Amount		Hourly	Daily
Skwk	Skilled Workers Average (35 trades)	$20.80	$166.40	10.4%	14.6%	12.6%	10%	47.6%	$ 9.90		$30.70	$245.60
	Helpers Average (5 trades)	15.90	127.20	11.1		12.7		48.4	7.70		23.60	188.80
	Foremen Average, Inside (50¢ over trade)	21.30	170.40	10.4		12.6		47.6	10.15		31.45	251.60
	Foremen Average, Outside ($2.00 over trade)	22.80	182.40	10.4		12.6		47.6	10.85		33.65	269.20
Clab	Common Building Laborers	16.10	128.80	11.4		10.8		46.8	7.55		23.65	189.20
Asbe	Asbestos Workers	23.00	184.00	8.8		15.7		49.1	11.30		34.30	274.40
Boil	Boilermakers	23.00	184.00	7.2		16.0		47.8	11.00		34.00	272.00
Bric	Bricklayers	20.55	164.40	8.5		10.7		43.8	9.00		29.55	236.40
Brhe	Bricklayer Helpers	16.15	128.40	8.5		10.7		43.8	7.05		23.20	185.60
Carp	Carpenters	20.55	164.40	11.4		10.8		46.8	9.60		30.15	241.20
Cefi	Cement Finishers	19.70	157.60	6.6		10.8		42.0	8.30		27.95	223.60
Elec	Electricians	22.65	181.20	4.5		15.9		45.0	10.20		32.85	262.80
Elev	Elevator Constructors	23.05	184.40	6.0		15.9		46.5	10.70		33.75	270.00
Eqhv	Equipment Operators, Crane or Shovel	21.20	169.60	7.8		13.8		46.2	9.80		31.00	248.00
Eqmd	Equipment Operators, Medium Equipment	20.75	166.00	7.8		13.8		46.2	9.60		30.35	242.80
Eqlt	Equipment Operators, Light Equipment	19.60	156.80	7.8		13.8		46.2	9.05		28.65	229.20
Eqol	Equipment Operators, Oilers	17.55	140.40	7.8		13.8		46.2	8.10		25.65	205.20
Eqmm	Equipment Operators, Master Mechanics	22.00	176.00	7.8		13.8		46.2	10.15		32.15	257.20
Glaz	Glaziers	20.75	166.00	8.7		10.8		44.1	9.15		29.90	239.20
Lath	Lathers	20.50	164.00	7.2		10.8		42.6	8.75		29.25	234.00
Marb	Marble Setters	20.25	162.00	8.5		10.7		43.8	8.85		29.10	232.80
Mill	Millwrights	21.25	170.00	7.2		10.8		42.6	9.05		30.30	242.40
Mstz	Mosaic and Terrazzo Workers	20.05	160.40	6.0		10.7		41.3	8.30		28.35	226.80
Pord	Painters, Ordinary	19.55	156.40	8.9		10.8		44.3	8.65		28.20	225.60
Psst	Painters, Structural Steel	20.20	161.60	29.4		10.1		64.1	12.95		33.15	265.20
Pape	Paper Hangers	19.75	158.00	8.9		10.8		44.3	8.75		28.50	228.00
Pile	Pile Drivers	20.35	162.80	18.3		15.7		58.6	11.95		32.30	258.40
Plas	Plasterers	20.25	162.00	8.7		10.9		44.2	8.95		29.20	233.60
Plah	Plasterer Helpers	16.65	133.20	8.7		10.9		44.2	7.35		24.00	192.00
Plum	Plumbers	23.00	184.00	5.3		15.9		45.8	10.55		33.55	268.40
Rodm	Rodmen (Reinforcing)	22.10	176.80	18.6		13.3		56.5	12.50		34.60	276.80
Rofc	Roofers, Composition	19.15	153.20	20.8		10.4		55.8	10.70		29.85	238.80
Rots	Roofers, Tile & Slate	19.25	154.00	20.8		10.4		55.8	10.75		30.00	240.00
Rohe	Roofer Helpers (Composition)	14.20	113.60	20.8		10.4		55.8	7.90		22.10	176.80
Shee	Sheet Metal Workers	23.10	184.80	7.0		15.8		47.4	10.95		34.05	272.40
Spri	Sprinkler Installers	23.85	190.80	6.1		15.9		46.6	11.10		34.95	279.60
Stpi	Steamfitters or Pipefitters	23.30	186.40	5.3		15.9		45.8	10.70		33.95	271.60
Ston	Stone Masons	20.60	164.80	8.5		10.7		43.8	9.00		29.60	236.80
Sswk	Structural Steel Workers	22.10	176.80	21.5		13.7		59.8	13.20		35.30	282.40
Tilf	Tile Layers (Floor)	20.00	160.00	6.0		10.7		41.3	8.25		28.25	226.00
Tilh	Tile Layer Helpers	16.10	128.80	6.0		10.7		41.3	6.65		22.75	182.00
Trlt	Truck Drivers, Light	16.70	133.60	9.8		10.7		45.1	7.55		24.25	194.00
Trhv	Truck Drivers, Heavy	16.95	135.60	9.8		10.7		45.1	7.65		24.60	196.80
Sswl	Welders, Structural Steel	22.10	176.80	21.5		13.7		59.8	13.20		35.30	282.40
Wrck	*Wrecking	16.10	128.80	23.1	↓	10.4	↓	58.1	9.35		25.45	203.60

*Not included in Averages.

Figure 20.1

Installing Contractor's Overhead & Profit

Below are the **average** installing contractor's percentage mark-ups applied to base labor rates to arrive at typical billing rates.

Column A: Labor rates are based on average wages for 7 major U.S. regions. Base rates including fringe benefits are listed hourly and daily. These figures are the sum of the wage rate and employer-paid fringe benefits such as vacation pay and employer-paid health costs.

Column B: Workers' Compensation rates are the national average of state rates established for each trade.

Column C: Column C lists average fixed overhead figures for all trades. Included are Federal and State Unemployment costs set at 5.5%; Social Security Taxes (FICA) set at 7.15%; Builder's Risk Insurance costs set at 1.14%; and Public Liability costs set at 0.82%. All the percentages except those for Social Security Taxes vary from state to state as well as from company to company.

Column D and E: Percentages in Columns D and E are based on the presumption that the installing contractor has annual billing of $500,000 and up. Overhead percentages may increase with smaller annual billing. The overhead percentages for any given contractor may vary greatly and depend on a number of factors, such as the contractor's annual volume, engineering and logistical support costs, and staff requirements. The figures for overhead and profit will also vary depending on the type of job, the job location, and the prevailing economic conditions. All factors should be examined very carefully for each job.

Column F: Column F lists the total of columns B, C, D, and E.

Column G: Column G is Column A (hourly base labor rate) multiplied by the percentage in Column F (O&P percentage).

Column H: Column H is the total of Column A (hourly base labor rate) plus Column G (Total O&P).

Column I: Column I is Column H multiplied by eight hours.

Abbr.	Trade	A Base Rate Incl. Fringes Hourly	A Base Rate Incl. Fringes Daily	B Workers' Comp. Ins.	C Average Fixed Overhead	D Subs Overhead	E Subs Profit	F Subs Total Overhead & Profit %	F Subs Total Overhead & Profit Amount	H Rate with Subs O & P Hourly	I Rate with Subs O & P Daily
Skwk	Skilled Workers Average	$12.30	$ 98.40	10.4%	14.6%	22.8%	10%	57.8%	$ 7.10	$19.40	$155.20
	Helpers Average ($2.00 under trade)	10.30	82.40	11.1		23.0		58.7	6.05	16.35	130.80
	Foremen Average, ($2.00 over trade)	14.30	114.40	10.4		22.8		57.8	8.25	22.55	180.40
Clab	Laborers	9.70	77.60	11.4		21.0		57.0	5.55	15.25	122.00
Asbe	Pipe or Duct Insulators	13.55	108.40	8.8		26.0		59.4	8.05	21.60	172.80
Boil	Boilermakers	13.55	108.40	7.2		26.0		57.8	7.85	21.35	170.80
Bric	Brick or Block Masons	12.10	96.80	8.5		21.0		54.1	6.55	18.65	149.20
Carp	Carpenters	12.20	97.60	11.4		21.0		57.0	6.95	19.15	153.20
Cefi	Cement Finishers	11.70	93.60	6.6		21.0		52.2	6.10	17.80	142.40
Elec	Electricians	13.30	106.40	4.5		26.0		55.1	7.35	20.65	165.20
Elev	Elevator Constructors	13.50	108.00	6.0		26.0		56.6	7.65	21.15	169.20
Eqhv	Equipment Operators, Crane	12.50	100.00	7.8		24.0		56.4	7.05	19.55	156.40
Eqmd	Equipment Operators	12.25	98.00	7.8		24.0		56.4	6.90	19.15	153.20
Eqmm	Equipment Mechanics	13.00	104.00	7.8		24.0		56.4	7.35	20.35	162.80
Glaz	Glaziers	12.30	98.40	8.7		21.0		54.3	6.70	19.00	152.00
Lath	Lathers	12.20	97.60	7.2		21.0		52.8	6.45	18.65	149.20
Mill	Millwrights	12.50	100.00	7.2		21.0		52.8	6.60	19.10	152.80
Pord	Painters	11.60	92.80	8.9		21.0		54.5	6.30	17.90	143.20
Pile	Pile Drivers	12.05	96.40	18.3		26.0		68.9	8.30	20.35	162.80
Plas	Plasterers	11.95	95.60	8.7		21.0		54.3	6.50	18.45	147.60
Plum	Plumbers	13.45	107.60	5.3		26.0		55.9	7.50	20.95	167.60
Rodm	Rodmen (Reinforcing)	13.00	104.00	18.6		24.0		67.2	8.75	21.75	174.00
Rofc	Roofers	11.40	91.20	20.8		21.0		66.4	7.55	18.95	151.60
Shee	Sheet Metal Workers	13.50	108.00	7.0		26.0		57.6	7.75	21.25	170.00
Spri	Sprinkler Installers	13.85	110.80	6.1		26.0		56.7	7.85	21.70	173.60
Stpi	Pipefitters	13.65	109.20	5.3		26.0		55.9	7.65	21.30	170.40
Ston	Stone Masons	12.20	97.60	8.5		21.0		54.1	6.60	18.80	150.40
Sswk	Structural Steel Erectors	13.00	104.00	21.5		24.0		70.1	9.10	22.10	176.80
Tilf	Flooring Installers	11.90	95.20	6.0		21.0		51.6	6.15	18.05	144.40
Trhv	Truck Drivers	10.20	81.60	9.8		21.0		55.4	5.65	15.85	126.80
Wrck	Wreckers	9.75	78.00	23.1	↓	21.0	↓	68.7	6.70	16.45	131.60

Figure 20.2

CREWS

Crew J-3

Crew No.	Bare Costs Hr.	Bare Costs Daily	Incl. Subs O&P Hr.	Incl. Subs O&P Daily	Cost Per Man-hour Bare Costs	Cost Per Man-hour Incl. O&P
1 Terrazzo Worker	$20.05	$160.40	$28.35	$226.80	$18.17	$25.70
1 Terrazzo Helper	16.30	130.40	23.05	184.40		
1 Terrazzo grinder, electric		31.20		34.30		
1 Terrazzo mixer		85.50		94.05	7.29	8.02
16 M.H., Daily Totals		$407.50		$539.55	$25.46	$33.72

Crew J-4

Crew No.	Bare Costs Hr.	Bare Costs Daily	Incl. Subs O&P Hr.	Incl. Subs O&P Daily	Cost Per Man-hour Bare Costs	Cost Per Man-hour Incl. O&P
1 Tile Layer	$20.00	$160.00	$28.25	$226.00	$18.05	$25.50
1 Tile Layer Helper	16.10	128.80	22.75	182.00		
16 M.H., Daily Totals		$288.80		$408.00	$18.05	$25.50

Crew K-1

Crew No.	Bare Costs Hr.	Bare Costs Daily	Incl. Subs O&P Hr.	Incl. Subs O&P Daily	Cost Per Man-hour Bare Costs	Cost Per Man-hour Incl. O&P
1 Carpenter	$20.55	$164.40	$30.15	$241.20	$18.62	$27.20
1 Truck Driver (light)	16.70	133.60	24.25	194.00		
1 Truck w/Power Equip.		178.85		196.70	11.17	12.29
16 M.H., Daily Totals		$476.85		$631.90	$29.79	$39.49

Crew K-2

Crew No.	Bare Costs Hr.	Bare Costs Daily	Incl. Subs O&P Hr.	Incl. Subs O&P Daily	Cost Per Man-hour Bare Costs	Cost Per Man-hour Incl. O&P
1 Struc. Steel Foreman	$24.10	$192.80	$38.50	$308.00	$20.96	$32.68
1 Struc. Steel Worker	22.10	176.80	35.30	282.40		
1 Truck Driver (light)	16.70	133.60	24.25	194.00		
1 Truck w/Power Equip.		178.85		196.70	7.45	8.19
24 M.H., Daily Totals		$682.05		$981.10	$28.41	$40.87

Crew L-1

Crew No.	Bare Costs Hr.	Bare Costs Daily	Incl. Subs O&P Hr.	Incl. Subs O&P Daily	Cost Per Man-hour Bare Costs	Cost Per Man-hour Incl. O&P
1 Electrician	$22.65	$181.20	$32.85	$262.80	$22.82	$33.20
1 Plumber	23.00	184.00	33.55	268.40		
16 M.H., Daily Totals		$365.20		$531.20	$22.82	$33.20

Crew L-2

Crew No.	Bare Costs Hr.	Bare Costs Daily	Incl. Subs O&P Hr.	Incl. Subs O&P Daily	Cost Per Man-hour Bare Costs	Cost Per Man-hour Incl. O&P
1 Carpenter	$20.55	$164.40	$30.15	$241.20	$18.22	$26.87
1 Helper	15.90	127.20	23.60	188.80		
16 M.H., Daily Totals		$291.60		$430.00	$18.22	$26.87

Crew L-3

Crew No.	Bare Costs Hr.	Bare Costs Daily	Incl. Subs O&P Hr.	Incl. Subs O&P Daily	Cost Per Man-hour Bare Costs	Cost Per Man-hour Incl. O&P
1 Carpenter	$20.55	$164.40	$30.15	$241.20	$21.71	$31.80
.5 Electrician	22.65	90.60	32.85	131.40		
.5 Sheet Metal Worker	23.10	92.40	34.05	136.20		
16 M.H., Daily Totals		$347.40		$508.80	$21.71	$31.80

Crew L-4

Crew No.	Bare Costs Hr.	Bare Costs Daily	Incl. Subs O&P Hr.	Incl. Subs O&P Daily	Cost Per Man-hour Bare Costs	Cost Per Man-hour Incl. O&P
2 Skilled Workers	$20.80	$332.80	$30.70	$491.20	$19.16	$28.33
1 Helper	15.90	127.20	23.60	188.80		
24 M.H., Daily Totals		$460.00		$680.00	$19.16	$28.33

Crew L-5

Crew No.	Bare Costs Hr.	Bare Costs Daily	Incl. Subs O&P Hr.	Incl. Subs O&P Daily	Cost Per Man-hour Bare Costs	Cost Per Man-hour Incl. O&P
1 Struc. Steel Foreman	$24.10	$192.80	$38.50	$308.00	$22.25	$35.14
5 Struc. Steel Workers	22.10	884.00	35.30	1412.00		
1 Equip. Oper. (crane)	21.20	169.60	31.00	248.00		
1 Hyd. Crane, 25 Ton		407.60		448.35	7.27	8.00
56 M.H., Daily Totals		$1654.00		$2416.35	$29.52	$43.14

Crew L-6

Crew No.	Bare Costs Hr.	Bare Costs Daily	Incl. Subs O&P Hr.	Incl. Subs O&P Daily	Cost Per Man-hour Bare Costs	Cost Per Man-hour Incl. O&P
1 Plumber	$23.00	$184.00	$33.55	$268.40	$22.88	$33.31
.5 Electrician	22.65	90.60	32.85	131.40		
12 M.H., Daily Totals		$274.60		$399.80	$22.88	$33.31

Crew L-7

Crew No.	Bare Costs Hr.	Bare Costs Daily	Incl. Subs O&P Hr.	Incl. Subs O&P Daily	Cost Per Man-hour Bare Costs	Cost Per Man-hour Incl. O&P
2 Carpenters	$20.55	$328.80	$30.15	$482.40	$19.57	$28.67
1 Building Laborer	16.10	128.80	23.65	189.20		
.5 Electrician	22.65	90.60	32.85	131.40		
28 M.H., Daily Totals		$548.20		$803.00	$19.57	$28.67

Crew L-8

Crew No.	Bare Costs Hr.	Bare Costs Daily	Incl. Subs O&P Hr.	Incl. Subs O&P Daily	Cost Per Man-hour Bare Costs	Cost Per Man-hour Incl. O&P
2 Carpenters	$20.55	$328.80	$30.15	$482.40	$21.04	$30.83
.5 Plumber	23.00	92.00	33.55	134.20		
20 M.H., Daily Totals		$420.80		$616.60	$21.04	$30.83

Crew L-9

Crew No.	Bare Costs Hr.	Bare Costs Daily	Incl. Subs O&P Hr.	Incl. Subs O&P Daily	Cost Per Man-hour Bare Costs	Cost Per Man-hour Incl. O&P
1 Labor Foreman (inside)	$16.60	$132.80	$24.35	$194.80	$18.27	$27.41
2 Building Laborers	16.10	257.60	23.65	378.40		
1 Struc. Steel Worker	22.10	176.80	35.30	282.40		
.5 Electrician	22.65	90.60	32.85	131.40		
36 M.H., Daily Totals		$657.80		$987.00	$18.27	$27.41

Crew M-1

Crew No.	Bare Costs Hr.	Bare Costs Daily	Incl. Subs O&P Hr.	Incl. Subs O&P Daily	Cost Per Man-hour Bare Costs	Cost Per Man-hour Incl. O&P
3 Elevator Constructors	$23.05	$553.20	$33.75	$810.00	$21.89	$32.06
1 Elevator Apprentice	18.44	147.52	27.00	216.00		
Hand Tools		55.00		60.50	1.71	1.89
32 M.H., Daily Totals		$755.72		$1086.50	$23.60	$33.95

Crew M-2

Crew No.	Bare Costs Hr.	Bare Costs Daily	Incl. Subs O&P Hr.	Incl. Subs O&P Daily	Cost Per Man-hour Bare Costs	Cost Per Man-hour Incl. O&P
2 Millwrights	$21.25	$340.00	$30.30	$484.80	$21.25	$30.30
Power Tools		21.20		23.30	1.32	1.45
16 M.H., Daily Totals		$361.20		$508.10	$22.57	$31.75

Crew Q-1

Crew No.	Bare Costs Hr.	Bare Costs Daily	Incl. Subs O&P Hr.	Incl. Subs O&P Daily	Cost Per Man-hour Bare Costs	Cost Per Man-hour Incl. O&P
1 Plumber	$23.00	$184.00	$33.55	$268.40	$20.70	$30.20
1 Plumber Apprentice	18.40	147.20	26.85	214.80		
16 M.H., Daily Totals		$331.20		$483.20	$20.70	$30.20

Crew Q-2

Crew No.	Bare Costs Hr.	Bare Costs Daily	Incl. Subs O&P Hr.	Incl. Subs O&P Daily	Cost Per Man-hour Bare Costs	Cost Per Man-hour Incl. O&P
2 Plumbers	$23.00	$368.00	$33.55	$536.80	$21.46	$31.31
1 Plumber Apprentice	18.40	147.20	26.85	214.80		
24 M.H., Daily Totals		$515.20		$751.60	$21.46	$31.31

Crew Q-3

Crew No.	Bare Costs Hr.	Bare Costs Daily	Incl. Subs O&P Hr.	Incl. Subs O&P Daily	Cost Per Man-hour Bare Costs	Cost Per Man-hour Incl. O&P
1 Plumber Foreman (ins)	$23.50	$188.00	$34.25	$274.00	$21.97	$32.05
2 Plumbers	23.00	368.00	33.55	536.80		
1 Plumber Apprentice	18.40	147.20	26.85	214.80		
32 M.H., Daily Totals		$703.20		$1025.60	$21.97	$32.05

Crew Q-4

Crew No.	Bare Costs Hr.	Bare Costs Daily	Incl. Subs O&P Hr.	Incl. Subs O&P Daily	Cost Per Man-hour Bare Costs	Cost Per Man-hour Incl. O&P
1 Plumber Foreman (ins)	$23.50	$188.00	$34.25	$274.00	$21.97	$32.05
1 Plumber	23.00	184.00	33.55	268.40		
1 Welder (plumber)	23.00	184.00	33.55	268.40		
1 Plumber Apprentice	18.40	147.20	26.85	214.80		
1 Electric Welding Mach.		19.00		20.90	.59	.65
32 M.H., Daily Totals		$722.20		$1046.50	$22.56	$32.70

Crew Q-5

Crew No.	Bare Costs Hr.	Bare Costs Daily	Incl. Subs O&P Hr.	Incl. Subs O&P Daily	Cost Per Man-hour Bare Costs	Cost Per Man-hour Incl. O&P
1 Steamfitter	$23.30	$186.40	$33.95	$271.60	$20.97	$30.57
1 Steamfitter Apprentice	18.64	149.12	27.20	217.60		
16 M.H., Daily Totals		$335.52		$489.20	$20.97	$30.57

Figure 20.3

Equipment

Construction equipment, when used for an interior project, can be a very expensive item. For example, for a sixth floor job, the elevator is too small to transport the drywall. A crane must be used, and a window removed, and other work essentially halted during delivery. This can cost thousands of dollars. Estimators must carefully address the issue of equipment and related expenses, when required. Equipment costs can be divided into the two following categories:

- *Rental, lease or ownership costs.* These costs may be determined based upon hourly, daily, weekly, monthly or annual increments. These fees or payments only buy the "right" to use the equipment (i.e., exclusive of operating costs).
- *Operating Costs.* Once the "right" of use is obtained, costs are incurred for actual use or operation. These costs may include fuel, lubrication, maintenance, and parts.

Equipment costs, as described above, do not include the labor expense of operators. However, some cost books and suppliers may include the operator in the quoted price for equipment as an "operated" rental cost. In other words, the equipment is priced as if it were a subcontract cost. Equipment ownership costs apply to both leased and owned equipment. The operating costs of equipment, whether rented, leased or owned, are available from the following sources (listed in order of reliability):

1. The company's own records
2. Annual cost books containing equipment operating costs, such as Means *Interior Cost Data*
3. Manufacturers' estimates
4. Text books dealing with equipment operating costs

These operating costs consist of fuel, lubrication, expendable parts replacement, minor maintenance, transportation, and mobilizing costs. For estimating purposes, the equipment ownership and operating costs should be listed separately. In this way, the decision to rent, subcontract or purchase can be decided project by project.

There are two commonly used methods for including equipment costs in a construction estimate. The first is to include the equipment as a part of the construction task for which it is used. In this case, costs are included in each line item as a separate unit price. The advantage of this method is that costs are allocated to the division or task that actually incurs the expense. As a result, more accurate records can be kept for each installed component. A disadvantage of this method occurs in the pricing of equipment that may be used for many different tasks. Duplication of costs can occur in this instance. Another disadvantage is that the budget may be left short for the following reason: the estimate may only reflect two hours for a crane truck, when the minimum cost of such a crane is usually a daily (8-hour) rental charge.

The second method for including equipment costs in the estimate is to keep all such costs separate and to include them in Division 1 as a part of Project Overhead. The advantage of this method is that all equipment costs are grouped together, and that machines that may be used for several tasks are included (without duplication). One disadvantage is that for future estimating purposes, equipment costs will be known only on a job basis and not per installed unit.

Whichever method is used, the estimator must be consistent, and must be sure that all equipment costs are included but not duplicated. The estimating method should be the same as that chosen for cost monitoring and accounting. In this way, the data will be available both for monitoring the project's costs and for bidding future projects.

Subcontractors

Subcontractors may account for a large percentage of an interiors project. When subcontractors are used, quotations should be solicited and analyzed in the same way as material quotes. A primary concern is that the bid covers the work according to the plans and specifications, and that all appropriate work alternates and allowances, if any, are included. Any exclusions should be clearly stated and explained. If the bid is received verbally, a form such as that shown in Chapter 19, Figure 19.1, will help to assure that it is documented accurately. Any unique scheduling or payment requirements must be noted and evaluated. Such requirements could affect (restrict or enhance) the normal progress of the project, and should therefore be known in advance.

The estimator should note how long the subcontract bid will be honored. This time period usually varies from 30 to 90 days and is often included as a condition in complete bids.

The estimator should know or verify the bonding capability and capacity of unfamiliar subcontractors when required. This act may be necessary when bidding in a new location. Other than word of mouth, these inquiries may be the only way to confirm subcontractor reliability.

Project Overhead

Project Overhead represents those construction costs that are usually included in Division One — General Requirements. Typical items are supervisory personnel, job engineers, cleanup, and temporary heat and power. While these items may not be directly part of the physical structure, they are a part of the project. Project Overhead, like all other direct costs, can be separated into material, labor, and equipment components. Figures 20.4a and 20.4b are examples of a form which can help ensure that all appropriate costs are included. This form is for general construction but could easily be adapted to the specific requirements of interior construction.

Some may not agree that certain items (such as equipment or scaffolding) should be included as Project Overhead, and might prefer to list such items in another division. Ultimately, it is not important *where* each item is incorporated into the estimate but that *every item is included somewhere*.

Project Overhead often includes time-related items; equipment rental, supervisory labor, and temporary utilities are examples. The cost for these items depends upon the duration of the project. A preliminary schedule should, therefore, be developed *prior* to completion of the estimate so that time-related items can be properly counted. This will be further discussed in Chapter 23, ''Pre-Bid Scheduling''.

🔺 Means Forms

PROJECT
OVERHEAD SUMMARY

PROJECT

SHEET NO.

ESTIMATE NO.

LOCATION   ARCHITECT   DATE

QUANTITIES BY: PRICES BY: EXTENSIONS BY: CHECKED BY:

DESCRIPTION	QUANTITY	UNIT	MATERIAL/EQUIPMENT		LABOR		TOTAL COST	
			UNIT	TOTAL	UNIT	TOTAL	UNIT	TOTAL
Job Organization: Superintendent								
Project Manager								
Timekeeper & Material Clerk								
Clerical								
Safety, Watchman & First Aid								
Travel Expense: Superintendent								
Project Manager								
Engineering: Layout								
Inspection/Quantities								
Drawings								
CPM Schedule								
Testing: Soil								
Materials								
Structural								
Equipment: Cranes								
Concrete Pump, Conveyor, Etc.								
Elevators, Hoists								
Freight & Hauling								
Loading, Unloading, Erecting, Etc.								
Maintenance								
Pumping								
Scaffolding								
Small Power Equipment/Tools								
Field Offices: Job Office								
Architect/Owner's Office								
Temporary Telephones								
Utilities								
Temporary Toilets								
Storage Areas & Sheds								
Temporary Utilities: Heat								
Light & Power								
Water								
PAGE TOTALS								

II.89

Page 1 of 2

Figure 20.4a

⚓ Means Forms

DESCRIPTION	QUANTITY	UNIT	MATERIAL/EQUIPMENT		LABOR		TOTAL COST	
			UNIT	TOTAL	UNIT	TOTAL	UNIT	TOTAL
Totals Brought Forward								
Winter Protection: Temp. Heat/Protection								
Snow Plowing								
Thawing Materials								
Temporary Roads								
Signs & Barricades: Site Sign								
Temporary Fences								
Temporary Stairs, Ladders & Floors								
Photographs								
Clean Up								
Dumpster								
Final Clean Up								
Punch List								
Permits: Building								
Misc.								
Insurance: Builders Risk								
Owner's Protective Liability								
Umbrella								
Unemployment Ins. & Social Security								
Taxes								
City Sales Tax								
State Sales Tax								
Bonds								
Performance								
Material & Equipment								
Main Office Expense								
Special Items								
TOTALS:								

II.91

Figure 20.4b

Bonds

Although bonds are really a type of "direct cost", they are priced and based upon the total "bid" or "selling price". For that reason, they are generally figured after indirect costs have been added. Bonding requirements for a project will be specified in Division 1 — General Requirements of the specifications, and will be included in the construction contract. Various types of bonds are discussed in the Introduction to this book.

Sales Tax

Sales tax varies from state to state and often from city to city within a state (see Figure 20.5). Larger cities may have a sales tax in addition to the state sales tax. Some localities also impose separate sales taxes on labor and equipment.

When bidding takes place in unfamiliar locations, the estimator should check with local agencies regarding the amount and the method of payment of sales tax. Local authorities may require owners to withhold payments to out-of-state contractors until payment of all required sales tax has been verified. Sales tax is often taken for granted or even omitted and, as can be seen in Figure 20.5, can be as much as 7.5% of material costs. Indeed, this can represent a significant portion of the project's total cost. Conversely, some clients and/or their projects may be tax exempt. If this fact is unknown to the estimator, a large dollar amount for sales tax might be needlessly included in a bid.

Sales Tax Percentages on Materials by State							
State	Tax	State	Tax	State	Tax	State	Tax
Alabama	4%	Illinois	5%	Montana	0%	Rhode Island	6%
Alaska	0	Indiana	5	Nebraska	3.5	South Carolina	5
Arizona	5	Iowa	4	Nevada	5.75	South Dakota	4
Arkansas	4	Kansas	3	New Hampshire	0	Tennessee	5.5
California	6	Kentucky	5	New Jersey	6	Texas	4
Colorado	3	Louisiana	4	New Mexico	3.75	Utah	5.5
Connecticut	7.5	Maine	5	New York	4	Vermont	4
Delaware	0	Maryland	5	North Carolina	3	Virginia	4
District of Columbia	6	Massachusetts	5	North Dakota	4	Washington	6.5
Florida	5	Michigan	4	Ohio	5.5	West Virginia	5
Georgia	3	Minnesota	6	Oklahoma	3	Wisconsin	5
Hawaii	4	Mississippi	6	Oregon	0	Wyoming	3
Idaho	4	Missouri	6.225	Pennsylvania	6	Average	4.25%

Figure 20.5

Chapter 21
INDIRECT COSTS

Indirect costs are the "costs of doing business". These expenses are sometimes referred to as "burden" to the project. Indirect costs may include certain fixed, or known, expenses and percentages, as well as costs which can be variable and subjectively determined. Government authorities require payment of certain taxes and insurance, usually based upon labor costs and determined by trade. These are a type of fixed indirect cost. Office overhead, if well understood and established, can also be considered as a relatively fixed percentage. Profit and contingencies, however, are more variable and subjective. These figures are often determined based on the judgement and discretion of the person responsible for the company's growth and success.

If the direct costs for the same project have been carefully determined, they should not vary significantly from one estimator to another. It is the indirect costs that are often most responsible for variations among bids.

The direct costs of a project must be itemized, tabulated, and totalled before the indirect costs can be applied to the estimate. Indirect costs include:

- Taxes and insurance
- Office or Operating Overhead
- Profit
- Contingencies

Taxes and Insurance

The taxes and insurance included as indirect costs are most often related to the costs of labor and/or the type of work. This category may include Worker's Compensation, Builder's Risk, and Public Liability insurance, as well as employer-paid social security tax and unemployment insurance. By law, the employer must pay these expenses. Rates are based on the type and salary of the employees, as well as the location and/or type of business.

Office or Operating Overhead

Office overhead, or the cost of doing business, is perhaps one of the main reasons why so many contractors are unable to realize a profit, or even to stay in business. This is manifested in two ways. Either a company does not know its true overhead cost and, therefore, fails to mark up its costs enough to recover them; or management does not restrain or control overhead costs effectively and fails to remain competitive.

If a contractor does not know the costs of operating the business, then, more than likely, these costs will not be recovered. Many companies survive, and even turn a profit, by simply adding an arbitrary percentage for overhead to each job, without knowing how the percentage is derived or what is included. When annual volume changes significantly, whether by increase or decrease, the previously used percentage for overhead may no longer be valid. When such a volume change occurs, the owner often finds that the company is not doing as well as before and cannot determine the reasons. Chances are, overhead costs are not being fully recovered. The following lists items which may be included when determining office overhead costs:

Owner: This should include only a reasonable base salary and does not include profits. An owner's salary is *not* a company's profit.

Designer/Estimator: Since the owner may be primarily on the road getting business, this is the person who runs the daily operation of the company and is responsible for estimating. In some operations, the estimator who successfully wins a bid, then becomes the "project manager" and is responsible to the owner for its profitability.

Secretary/Receptionist: This person manages office operations and handles paperwork. A talented individual in this position can be a tremendous asset.

Office Staff Insurance & Taxes: These costs are for main office personnel and should include, but are not limited to, the following:
- Worker's Compensation
- FICA
- Unemployment
- Medical & other insurance
- Profit sharing, pension, etc.

Physical Plant Expenses: Whether the office, warehouse and/or yard are rented or owned, roughly the same costs are incurred. Telephone and utility costs will vary depending on the size of the building and the type of business. Office equipment includes items such as the rental of a copy machine and typewriters.

Professional Services: Accountant fees are primarily for quarterly audits. Legal fees go towards collecting receivables and contract disputes. In addition, a prudent contractor will have *every* contract read by his lawyer prior to signing.

Miscellaneous: There are many expenses that could be placed in this category. Association dues, seminars, travel, and entertainment may be included. Advertising includes the Yellow Pages, promotional materials, etc.

Uncollected Receivables: This amount can vary greatly, and is often affected by the overall economic climate. Depending upon the timing of "uncollectables", cash flow can be severely restricted and can cause serious financial problems, even for large companies. Sound cash planning and anticipation of such possibilities can help to prevent severe repercussions.

In order for a company to stay in business without losses (profit is not yet a factor), not only must all direct construction costs be paid, but all additional overhead costs must be recovered during the year in order to operate the office. The most common method for recovering these costs is to apply this percentage to each job over the course of the year. The percentage may be calculated and applied in two ways:

- Office overhead applied as a percentage of *labor costs only*. This method requires that labor and material costs be estimated separately.
- Office overhead applied as a percentage of *total project costs*. This is appropriate where material and labor costs are not estimated separately.

The estimator must also remember that, if volume changes significantly, then the percentage for office overhead should be recalculated for current conditions. The same is true if there are changes in office staff. Salaries are the major portion of office overhead costs. It should be noted that an additional percentage is commonly applied to material costs, for handling, regardless of the method of recovering office overhead costs. This percentage is more easily calculated if material costs are estimated and listed separately.

Profit

Determining a fair and reasonable percentage to be included for profit is not an easy task. This responsibility is usually left to the owner or chief estimator. Experience is crucial in anticipating what profit the market will bear. The economic climate, competition, knowledge of the project, and familiarity with the architect, designer, or owner all affect the way in which profit is determined.

Contingencies

Like profit, contingencies can also be difficult to quantify. Especially appropriate in preliminary budgets, the addition of a contingency is meant to protect the contractor as well as to give the owner a realistic estimate of potential project costs.

A contingency percentage should be based on the number of "unknowns" in a project, or the level of risk involved. This percentage should be inversely proportional to the amount of planning detail that has been done for the project. If complete plans and specifications are supplied, and the estimate is thorough and precise, then there is little need for a contingency. Figure 21.1, from Means *Interior Cost Data* 1987, lists suggested contingency percentages that may be added to an estimate based on the stage of planning and development.

If an estimate is priced and each individual item is rounded up, or "padded", this is, in essence, adding a contingency. This method can cause problems, however, because the estimator can never be quite sure of what is the actual cost and what is the "padding", or safety margin, for each item. At the summary, the estimator cannot determine exactly how much has been included as a contingency factor for the project as a whole. A much more accurate and controllable approach is to price the estimate precisely and then add one contingency amount at the bottom line.

		1.1 Overhead	CREW	DAILY OUTPUT	MAN-HOURS	UNIT	BARE COSTS				TOTAL INCL O&P	
							MAT.	LABOR	EQUIP.	TOTAL		
020	0011	ARCHITECTURAL FEES ⑩										020
	0020	For work to $10,000				Project					15%	
	0040	To $25,000										
	0060	To $100,000				Project					10%	
	0080	To $500,000									10%	
	0090	To $1,000,000				↓					7%	
040	0010	CLEANING UP After job completion, allow				Job					.30%	040
	0031	Rubbish removal, see division 2.1-430										
	0050	Cleanup of floor area, continuous, per day	A-5	12	1.500	M.S.F.	1.50	24	1.20	26.70	39	
	0100	Final	"	11.50	1.570	"	1.60	25	1.25	27.85	40	
060	0011	CONSTRUCTION COST INDEX For 162 major U.S. and										060
	0020	Canadian cities, total cost, min. (Greensboro, NC)				%					80.40%	
	0050	Average									100%	
	0100	Maximum (Anchorage, AK)				↓					132.60%	
090	0010	CONSTRUCTION MANAGEMENT FEES $1,000,000 job, minimum				Project					4.50%	090
	0050	Maximum									7.50%	
110	0010	CONTINGENCIES Allowance to add at conceptual stage									15%	110
	0050	Schematic stage									10%	
	0100	Preliminary working drawing stage									7%	
	0150	Final working drawing stage				↓					2%	
120	0014	CONTRACTOR EQUIPMENT See division 1.5										120
140	0010	CREWS For building construction, see foreword										140
150	0010	ENGINEERING FEES Educational planning consultant, minimum				Project					.50%	150
	0100	Maximum				"					2.50%	
	0200	Electrical, minimum				Contrct					4.10%	
	0300	Maximum									10.10%	
	0600	Food service & kitchen equipment, minimum									8%	
	0700	Maximum									12%	
	1000	Mechanical (plumbing & HVAC), minimum									4.10%	
	1100	Maximum				↓					10.10%	
180	0010	INSURANCE Builders risk, standard, minimum ②				Job					.19%	180
	0050	Maximum									1.14%	
	0200	All-risk type, minimum									.20%	
	0250	Maximum									1.16%	
	0600	Public liability, average				↓					.82%	
	0610											
	0800	Workers' compensation & employer's liability, average ⑦										
	0850	by trade, carpentry, general				Payroll		11.41%				
	0900	Clerical						.42%				
	0950	Concrete						10.30%				
	1000	Electrical						4.46%				
	1050	Excavation						7.81%				
	1100	Glazing						8.73%				
	1150	Insulation						8.80%				
	1200	Lathing						7.21%				
	1250	Masonry						8.53%				
	1300	Painting & decorating						8.91%				
	1350	Pile driving						18.33%				
	1400	Plastering						8.73%				
	1450	Plumbing						5.31%				
	1500	Roofing						20.77%				
	1550	Sheet metal work (HVAC)						7%				
	1600	Steel erection, structural						21.51%				
	1650	Tile work, interior ceramic						5.96%				
	1700	Waterproofing, brush or hand caulking						5.09%				
	1800	Wrecking						23.08%				
	2000	Range of 36 trades in 50 states, excl. wrecking, minimum						1.10%				
	2100	Average				↓		10.42%				

For expanded coverage of these items see *Means Building Construction Cost Data 1987*   1

Figure 21.1

Chapter 22

THE ESTIMATE SUMMARY

At the pricing stage of the estimate, there is typically a large amount of paperwork that must be assembled, analyzed, and tabulated. Generally, the information contained in this paperwork could be recorded on any or all of the following major categories:

- Quantity Takeoff sheets for all work items (Figure 18.1)
- Material supplier's written quotations
- Equipment or material supplier's or subcontractor's telephone quotations (Figure 19.1)
- Subcontractor's written quotations
- Equipment supplier's quotations
- Pricing sheets (Figures 18.2 and 18.3)
- Estimate summary sheets

In the "real world" of estimating, many quotations, especially for large material purchases and for subcontracts, are not received until the last minute before the bidding deadline. Therefore, a system is needed to efficiently handle the paperwork and to ensure that everything will get transferred once (and only once) from the quantity takeoff to the cost analysis sheets. Some general rules for this process are as follows:

- Write on only one side of any document, where possible.
- Use Telephone Quotation forms for uniformity in recording prices received by phone.
- Document the source of every quantity and price.
- Keep the entire estimate in one or more compartmentalized folders.
- If you are pricing your own materials, number and code each takeoff sheet and each pricing extension sheet as it is created. At the same time, keep an index list of each sheet by number. If a sheet is to be abandoned, write "VOID" on it, but do not discard it. Keep it until the bid is accepted to be able to account for all pages and sheets.

All subcontract costs should be properly noted and listed separately. These costs contain the subcontractor's markups and may be treated differently from other direct costs when the estimator calculates the prime contractor's overhead and profit.

After all the unit prices and allowances have been entered on the pricing sheets, the costs are extended. In making the extensions, ignore the cents column and round all totals to the nearest dollar. In a column of figures, the cents will average out and will not be of consequence. Finally, each subdivision is added and the results checked, preferably by someone other than the person doing the extensions.

It is important to check the larger items for order of magnitude errors. If the total costs are divided by the floor area, the resulting square foot cost figures can be used to quickly pinpoint areas that are out of line with expected square foot costs. These cost figures should be recorded for comparison to past projects and as a resource for future estimating.

The takeoff and pricing method, as discussed, has been to utilize a Quantity Sheet for the material takeoff, and to transfer the data to a Cost Analysis form for pricing the material, labor, and equipment items.

An alternative to this method is a consolidation of the takeoff task and pricing on a single form. This approach works well for smaller bids and for change orders. An example, the Consolidated Estimate form, is shown in Figure 22.1. The same sequences and recommendations used to complete the Quantity Sheet and Cost Analysis form are to be followed when using the Consolidated Estimate form to price the estimate.

When the pricing of all direct costs is complete, the estimator has two choices: to make all further price changes and adjustments on the Cost Analysis or Consolidated Estimate sheets, *or* to transfer the total costs for each subdivision to an Estimate Summary sheet so that all further price changes, until bid time, will be done on one sheet. Any indirect cost markups and burdens will be figured on this sheet. An example of an Estimate Summary is shown in Figure 22.2.

Unless the estimate has a limited number of items, it is recommended that costs be transferred to an Estimate Summary sheet. This step should be double-checked since an error of transposition may easily occur. Pre-printed forms can be useful, although a plain columnar form may suffice. This summary with page numbers from each extension sheet can also serve as an index.

A company that repeatedly uses certain standard listings can save valuable time by having a custom Estimate Summary sheet printed with these items listed.The printed CSI division and subdivision headings serve as another type of checklist, ensuring that all required costs are included. Appropriate column headings or categories for any estimate summary form could be as follows:

- Material
- Labor
- Equipment
- Subcontractor
- Total

As items are listed in the proper columns, each category is added and appropriate markups applied to the total dollar values. Different percentages may be added to the sum of each column at the estimate summary. These percentages may include the following items, as discussed in Chapter 21:

- Taxes and Insurance
- Overhead
- Profit
- Contingencies

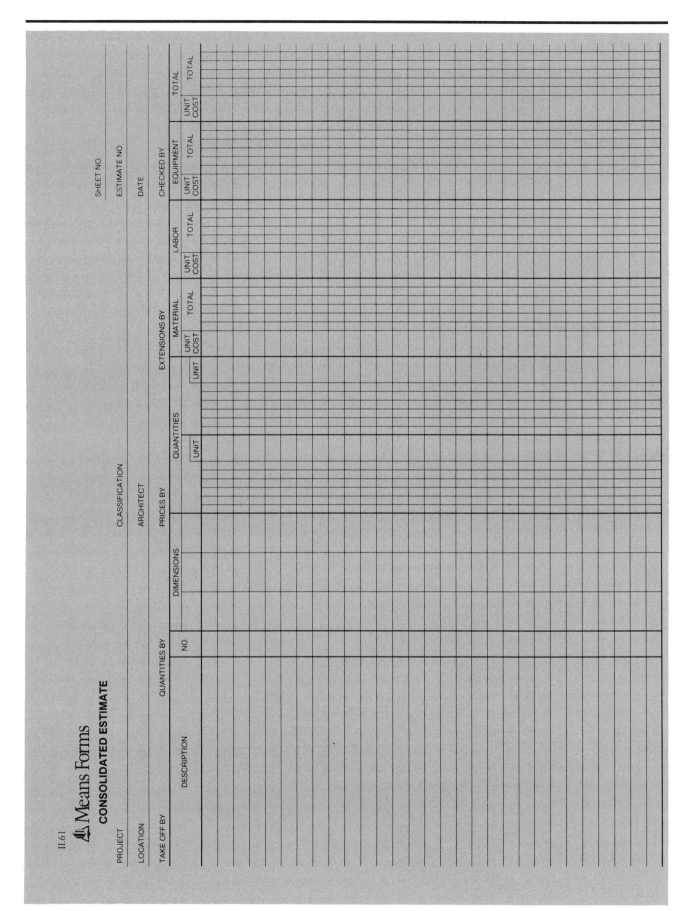

Figure 22.1

II.61

Means Forms

CONDENSED ESTIMATE SUMMARY

PROJECT				SHEET NO.	
LOCATION		TOTAL AREA/VOLUME		ESTIMATE NO.	
ARCHITECT		COST PER S.F./C.F.		DATE	
PRICES BY:		EXTENSIONS BY:		NO. OF STORIES	
				CHECKED BY:	

DIV.	DESCRIPTION	MATERIAL	LABOR	EQUIPMENT	SUBCONTRACT	TOTAL
1.0	General Requirements					
2.0	Site Work					
3.0	Concrete					
4.0	Masonry					
5.0	Metals					
6.0	Carpentry					
7.0	Moisture & Thermal Protection					
8.0	Doors, Windows, Glass					
9.0	Finishes					
10.0	Specialties					
11.0	Equipment					
12.0	Furnishings					
13.0	Special Construction					
14.0	Conveying Systems					
15.0	Mechanical					
16.0	Electrical					
	Subtotals					
	Sales Tax %					
	Overhead %					
	Subtotal					
	Profit %					
	Contingency %					
	Adjustments					
	TOTAL BID					

II.139

Figure 22.2

206

Chapter 23
PRE-BID
SCHEDULING

The need for planning and scheduling is clear once the contract is signed and work commences on the project. However, some scheduling is also important during the bidding stage for the following reasons:

- To determine if the project can be completed in the allotted or specified time using normal crew sizes
- To identify potential overtime requirements
- To determine the time requirements for supervision
- To anticipate possible temporary heat and power requirements
- To price certain general requirement items and overhead costs
- To budget for equipment usage
- To anticipate and justify material and equipment delivery requirements

The schedule produced prior to bidding may be a simple bar chart or network diagram that includes overall quantities, probable delivery times and available manpower. Network scheduling methods, such as the Critical Path Method (CPM) and the Precedence Chart simplify pre-bid scheduling because they do not require time-scaled line diagrams.

In the CPM Diagram, the activity is represented by an arrow. Nodes indicate start/stop between activities. The Precedence Diagram, on the other hand, shows the activity as a node with arrows used to denote precedence relationships between the activities. The precedence arrows may be used in different configurations to represent the sequential relationships between activities. Examples of CPM and Precedence diagrams are shown in Figures 23.1 and 23.2, respectively. In both systems, duration times are indicated along each path. The sequence (path) of activities requiring the most total time represents the shortest possible time (critical path) in which those activities may be completed.

For example, in both Figures 23.1 and 23.2, activities A, B, and C require 20 successive days for completion before activity G can begin. Activity paths for D and E (15 days), and for F (12 days) are shorter and can easily be completed during the 20-day sequence. Therefore, this 20-day sequence is the shortest possible time (i.e., the "critical path") for the completion of these activities – before activity G can begin.

Past experience or a prepared rough schedule may suggest that the time specified in the bidding documents is insufficient to complete the required work. In such cases, a more comprehensive schedule should be produced prior to bidding; this schedule will help to determine the added overtime or premium time work costs required to meet the completion date.

A three thousand square foot office renovation is used for one of the Sample Estimates in Part Three of this book. A preliminary schedule is needed to determine the supervision and manning requirements of the job.

A rough schedule for the interior work might be produced as shown in Figure 23.3. The man-days used to develop this schedule are derived from output figures determined in the estimate. Output can be determined based on the figures in Means *Interior Cost Data*. Man-days can also be figured by dividing the total labor cost shown on the estimate by the cost per man-day for each appropriate tradesperson.

As shown, the preliminary schedule can be used to determine supervision requirements, to develop appropriate crew sizes, and as a basis for ordering materials. All of these factors must be considered at this preliminary stage in order to determine how to meet the required completion date.

A pre-bid schedule can provide much more information than simple job duration. It can be used to refine the estimate by introducing realistic manpower projections. The schedule may also help the contractor to adjust the structure and size of the company based on projected requirements for months, even years ahead. A schedule can also become an effective tool for negotiating contracts.

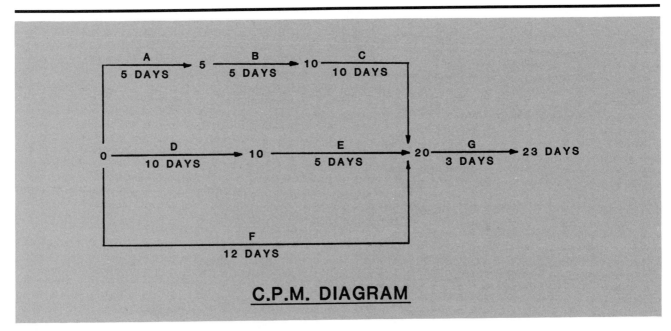

C.P.M. DIAGRAM

Figure 23.1

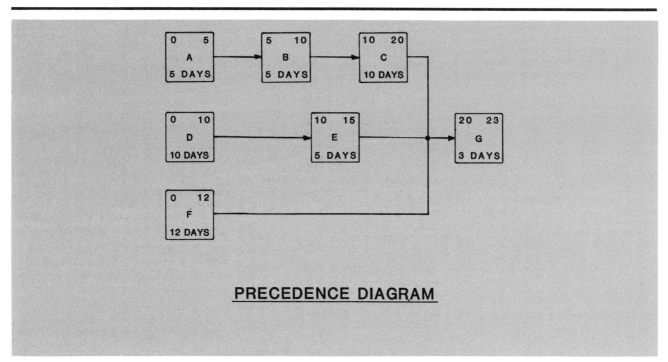

PRECEDENCE DIAGRAM

Figure 23.2

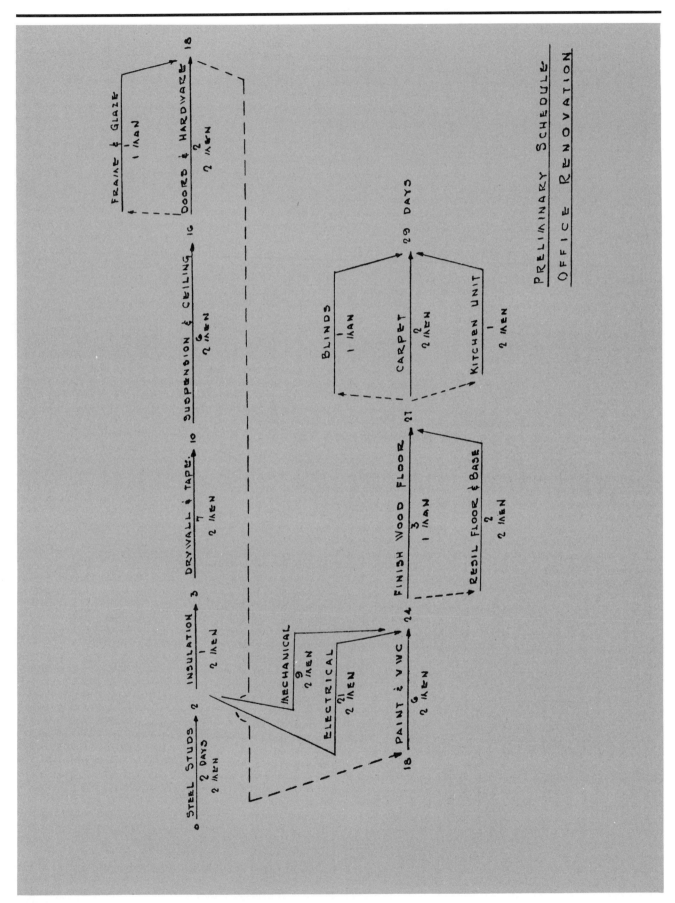

PRELIMINARY SCHEDULE
OFFICE RENOVATION

Figure 23.3

Part III
SAMPLE ESTIMATES

Chapter 24
USING MEANS INTERIOR COST DATA

Users of Means *Interior Cost Data* are chiefly interested in obtaining quick, reasonable, average prices for interior construction items. This is the primary purpose of the annual book – to eliminate guesswork when pricing unknowns. Many persons use the cost data, whether for bids or verification of quotations or budgets, without being fully aware of how the prices are obtained and derived. Without this knowledge, this resource is not being used to the fullest advantage. In addition to the basic cost data, the book also contains a wealth of information to aid the estimator, the contractor, the designer, and the owner to better plan and manage interior construction projects. Productivity data is provided in order to assist with scheduling. National labor rates are analyzed. Tables and charts for location and time adjustments are included and help the estimator tailor the prices to a specific location. The costs in Means *Interior Cost Data* consist of over 13,000 unit price line items, as well as prices for over 3500 interior systems. The unit price information, organized according to the Construction Specification Institute Masterformat Divisions, also provides an invaluable check list to the construction professional to assure that all required items are included in a project.

Format and Data

The major portion of Means *Interior Cost Data* is the Unit Price section. This is the primary source of unit cost data and is organized according to the CSI division index. This index was developed by representatives of all parties concerned with the building construction industry and has been accepted by the American Institute of Architects (AIA), the Associated General Contractors of America, Inc. (AGC) and the Construction Specifications Institute, Inc. (CSI).

CSI Masterformat Divisions:

Division 1 – General Requirements
Division 2 – Site Work
Division 3 – Concrete
Division 4 – Masonry
Division 5 – Metals
Division 6 – Wood & Plastics
Division 7 – Moisture-Thermal Control
Division 8 – Doors, Windows & Glass
Division 9 – Finishes

Division 10 – Specialties
Division 11 – Equipment
Division 12 – Furnishings
Division 13 – Special Construction
Division 14 – Conveying Systems
Division 15 – Mechanical
Division 16 – Electrical

Section B contains tables and reference charts. It also provides estimating procedures and explanations of cost development which support and supplement the unit price and systems cost data.

Section C, Assemblies Cost Tables, contains over 3,500 costs for related assemblies, or systems. Components of the systems are fully detailed and accompanied by illustrations.

Section D includes City Cost Indexes, representing the compilation of construction data for 162 major U. S. and Canadian cities. Cost factors are given for each city, by trade, relative to the national average. A comprehensive index and list of abbreviations is also included.

The prices presented in Means *Interior Cost Data* are national averages. Material and equipment costs are developed through annual contact with manufacturers, dealers, distributors and contractors throughout the United States. Means' staff of engineers is constantly updating prices and keeping abreast of changes and fluctuations within the industry. Labor rates are the national average of each trade as determined from union agreements from thirty major U. S. cities. Throughout the calendar year, as new wage agreements are negotiated, labor costs should be factored accordingly.

Following is a list of factors and assumptions on which the costs presented in Means *Interior Cost Data* have been based:

- *Quality*: The costs are based on methods, materials and workmanship in accordance with U.S. Government standards and represent good, sound construction practice.
- *Overtime*: The costs as presented, include *no* allowance for overtime. If overtime or premium time is anticipated, labor costs must be factored accordingly.
- *Productivity*: The daily output and man-hour figures are based on an eight hour workday, during daylight hours. The chart in Figure 24.1 shows that as the number of hours worked per day (over eight) increases, and as the days per week (over five) increase, production efficiency decreases.
- *Size of Project*: Costs in *Interior Cost Data* are based on interior projects which cost $20,000 and up. Large residential projects are also included.
- *Local Factors*: Weather conditions, season of the year, local union restrictions, and unusual building code requirements can all have a significant impact on construction costs. The availability of a skilled labor force, sufficient materials and even adequate energy and utilities will also affect costs. These factors vary in impact and are not necessarily dependent upon location. They must be reviewed for each project in every area.

In presenting prices in Means *Interior Cost Data* certain rounding rules are employed to make the numbers easy to use without significantly affecting accuracy. The rules are used consistently and are as follow:

Prices From	To	Rounded to Nearest
$ 0.01	$ 5.00	0.01
5.01	20.00	0.05
20.01	100.00	1.00
100.01	1,000.00	5.00
1,000.01	10,000.00	25.00
10,000.01	50,000.00	100.00
50,000.01	up	500.00

Section A – Unit Price Costs

The Unit Price section of *Interior Cost Data* contains a great deal of information in addition to the unit cost for each construction component. Figure 24.2 is a typical page, showing costs for drywall. Note that prices are included for several types of drywall, each based on type of insulation and finishing requirements. In addition, appropriate crews, workers and productivity data are indicated. The information and cost data is broken down and itemized in this way to provide for the most detailed pricing possible.

Days per Week	Hours per Day	Production Efficiency					Payroll Cost Factors	
		1 Week	2 Weeks	3 Weeks	4 Weeks	Average 4 Weeks	@ 1½ Times	@ 2 Times
5	8	100%	100%	100%	100%	100%	100%	100%
	9	100	100	95	90	96.25	105.6	111.1
	10	100	95	90	85	91.25	110.0	120.0
	11	95	90	75	65	81.25	113.6	127.3
	12	90	85	70	60	76.25	116.7	133.3
6	8	100	100	95	90	96.25	108.3	116.7
	9	100	95	90	85	92.50	113.0	125.9
	10	95	90	85	80	87.50	116.7	133.3
	11	95	85	70	65	78.75	119.7	139.4
	12	90	80	65	60	73.75	122.2	144.4
7	8	100	95	85	75	88.75	114.3	128.6
	9	95	90	80	70	83.75	118.3	136.5
	10	90	85	75	65	78.75	121.4	142.9
	11	85	80	65	60	72.50	124.0	148.1
	12	85	75	60	55	68.75	126.2	152.4

Figure 24.1

9.2 Drywall		CREW	DAILY OUTPUT	MAN-HOURS	UNIT	BARE COSTS				TOTAL INCL O&P
						MAT.	LABOR	EQUIP.	TOTAL	
020	1202	Studs and runners for partitions, see also 9.1-160									020
	1210										
	1500	Z bar, galvanized steel, 1-½" wide ⑩⑧	1 Carp	2.60	3.080	C.L.F.	20	63		83	115
	1600	2" wide	"	2.55	3.140	"	22	64		86	120
070	0010	DRYWALL Gypsum plasterboard, nailed or screwed to studs,									070
	0100	unless otherwise noted									
	0150	⅜" thick, on walls, standard, no finish included	2 Carp	2,000	.008	S.F.	.24	.16		.40	.51
	0200	On ceilings, standard, no finish included		1,400	.011		.24	.23		.47	.61
	0250	On beams, columns, or soffits, no finish included		750	.021		.24	.44		.68	.91
	0300	½" thick, on walls, standard, no finish included		1,800	.009		.25	.18		.43	.54
	0350	Taped and finished		965	.017		.27	.34		.61	.80
	0400	Fire resistant, no finish included		1,800	.009		.28	.18		.46	.58
	0450	Taped and finished		965	.017		.30	.34		.64	.83
	0500	Water resistant, no finish included		1,800	.009		.31	.18		.49	.61
	0550	Taped and finished	↓	965	.017	↓	.33	.34		.67	.86
	0650										
	1000	On ceilings, standard, no finish included	2 Carp	1,265	.013	S.F.	.25	.26		.51	.66
	1050	Taped and finished		765	.021		.27	.43		.70	.93
	1100	Fire resistant, no finish included		1,265	.013		.28	.26		.54	.69
	1150	Taped and finished		765	.021		.30	.43		.73	.96
	1200	Water resistant, no finish included		1,265	.013		.31	.26		.57	.72
	1250	Taped and finished		765	.021		.33	.43		.76	.99
	1500	On beams, columns, or soffits, standard, no finish included		675	.024		.25	.49		.74	.99
	1550	Taped and finished		475	.034		.27	.69		.96	1.31
	1600	Fire resistant, no finish included		675	.024		.28	.49		.77	1.02
	1650	Taped and finished		475	.034		.30	.69		.99	1.35
	1700	Water resistant, no finish included		675	.024		.31	.49		.80	1.06
	1750	Taped and finished		475	.034		.33	.69		1.02	1.38
	2000	⅝" thick, on walls, standard, no finish included		1,700	.009		.26	.19		.45	.57
	2050	Taped and finished		940	.017		.28	.35		.63	.82
	2100	Fire resistant, no finish included		1,700	.009		.29	.19		.48	.60
	2150	Taped and finished		940	.017		.31	.35		.66	.85
	2200	Water resistant, no finish included		1,700	.009		.34	.19		.53	.66
	2250	Taped and finished	↓	940	.017	↓	.36	.35		.71	.91
	2350										
	3000	On ceilings, standard, no finish included	2 Carp	1,175	.014	S.F.	.28	.28		.56	.72
	3050	Taped and finished		730	.022		.30	.45		.75	.99
	3100	Fire resistant, no finish included		1,175	.014		.29	.28		.57	.73
	3150	Taped and finished		730	.022		.31	.45		.76	1.00
	3200	Water resistant, no finish included		1,175	.014		.34	.28		.62	.78
	3250	Taped and finished		730	.022		.36	.45		.81	1.06
	3500	On beams, columns, or soffits, standard, no finish included		650	.025		.28	.51		.79	1.05
	3550	Taped and finished		450	.036		.30	.73		1.03	1.40
	3600	Fire resistant, no finish included		650	.025		.29	.51		.80	1.06
	3650	Taped and finished		450	.036		.31	.73		1.04	1.41
	3700	Water resistant, no finish included		650	.025		.34	.51		.85	1.12
	3750	Taped and finished		450	.036		.36	.73		1.09	1.47
	4000	Fireproofing, beams or columns, 2 layers, ½" thick		330	.048		.36	1		1.36	1.86
	4050	⅝" thick		300	.053		.44	1.10		1.54	2.09
	4100	3 layers, ½" thick		225	.071		.54	1.46		2	2.74
	4150	⅝" thick		210	.076		.66	1.57		2.23	3.02
	4600	Blueboard, ½" thick, standard, not incl. skim coat		1,800	.009		.25	.18		.43	.54
	4650	Fireproof		1,800	.009		.28	.18		.46	.58
	4700	⅝" thick, fireproof	↓	1,700	.009	↓	.29	.19		.48	.60
	5000	Sheathing, gypsum, see division 6.1-70									
	5050	For 1" thick coreboard on columns	2 Carp	480	.033	S.F.	.36	.69		1.05	1.40
	5100	For foil-backed board, add					.11			.11	.12
	5200	For high ceilings, over 8' high, add	2 Carp	3,060	.005		.10	.11		.21	.27
	5250	For prime coat (residential construction), add	↓	2,400	.007	↓	.05	.14		.19	.26
	5270	For textured spray, add		1,450	.011		.10	.23		.33	.44

105

Figure 24.2

Within each individual line item, there is a description of the construction component, information regarding typical crews designated to perform the work, and productivity shown as daily output and as man-hours. Costs are presented as "bare", or unburdened, as well as with mark-ups for overhead and profit. Figure 24.3 is a graphic representation of how to use the Unit Price section as presented in Means *Interior Cost Data.*

Line Numbers

Every construction item in the Means unit price cost data books has a unique line number. This line number acts as an address so that each item can be quickly located and/or referenced. The numbering system is based on the CSI Masterformat classification by division. In Figure 24.2, note the bold number in reverse type, "9.2". This number represents the major subdivision, in this case "Drywall", of the major CSI Division 9 – Finishes. All 16 divisions are organized in this manner. Within each subdivision, the data is broken down into major classifications. These major classifications are listed alphabetically and are designated by bold type for both numbers and descriptions. Each item, or line, is further defined by an individual number. As shown in Figure 24.3, the full line number for each item consists of: a major CSI subdivision number – a major classification number – an item line number. Each full line number describes a unique construction element. For example, in Figure 24.2, the line number for 1/2" drywall, on walls, taped and finished, is 9.2-070-0350.

Line Description

Each line has a text description of the item for which costs are listed. The description may be self-contained and all inclusive or, if indented, the complete description for a line is dependent upon the information provided above. All indented items are delineations (by size, color, material, etc.) or breakdowns of previously described items. An index is provided in the back of Means *Interior Cost Data* to aid in locating particular items.

Crew

For each construction element, (each line item), a minimum typical crew is designated as appropriate to perform the work. The crew may include one or more trades, foremen, craftsmen and helpers, and any equipment required for proper installation of the described item. If an individual trade installs the item using only hand tools, the smallest efficient number of tradesmen will be indicated (1 Carp, 2 Elec, etc.). Abbreviations for trades are shown in Figure 24.4. If more than one trade is required to install the item and/or if powered equipment is needed, a crew number will be designated (B-5, D-3, etc.). A complete listing of crews is presented in the Foreword pages of Means *Interior Cost Data* (see Figure 20.3). On these pages, each crew is broken down into the following components:

1. Number and type of workers designated.
2. Number, size, and type of any equipment required.
3. Hourly labor costs listed two ways: "bare" – base rate including fringe benefits; and including installing contractor's overhead and profit – "billing rate". (See Figure 24.4 from the inside back cover of Means *Interior Cost Data* for labor rate information).

HOW TO USE UNIT PRICE PAGES

Important
Prices in this section are listed in two ways: as bare costs and as costs including overhead and profit of the installing contractor. In most cases, if the work is to be subcontracted, it is best for a general contractor to add an additional 10% to the figures found in the column titled **"TOTAL INCL. O&P".**

Unit
The unit of measure listed here reflects the material being used in the line item. For example: a metal door is priced per each (Ea.)

Productivity
The daily output represents typical total daily amount of work that crew will produce. Man-hours are a unit-of measure for the labor involved in performing a task. To derive the total man-hours for a task, multiply the quantity of the item involved times the man-hour figure shown.

Line Number Determination
Major C.S.I. subdivision is **08.1** (two digits plus decimal point plus last digit)

Major classification within UCI subdivision is **210** (three digits)

Item line number is **0020** (four digits)

Complete line number is **08.1-210-0020**

Description
This line number identifies a commercial steel door, flush, hollow core, full panel, 20 ga., 2'-0" x 6'-8" to be installed at the rate of .800 man-hours per door. An F-2 crew can install 20 doors per day.

Crew F-2

Crew No.	Bare Costs		Incl. Subs O & P		Cost Per Man-hour	
Crew F-2	Hr.	Daily	Hr.	Daily	Bare Costs	Incl. O&P
2 Carpenters	$20.55	$328.80	$30.15	$482.40	$20.55	$30.15
Power Tools		21.20		23.30	1.32	1.45
16 M.H., Daily Totals		$350.00		$505.70	$21.87	$31.60

Bare Costs are developed as follows for line no. **08.1-210-0020**
Mat. is **Bare Material Cost ($108)**
Labor for Crew F2 = Man-hour Cost **($20.55)** × Man-hour Units **(.800)** = **$16.45**
Equip. for Crew F2 = Equip. Hour Cost **($1.32)** × Man-hour Units **(.800)** = **$1.06**
Total = **Mat. Cost ($108)** + **Labor Cost ($16.45)** + **Equip. Cost ($1.06)** = **$125.51** per door.
(**Note:** Equipment and Labor costs are derived from the Crew Tables. See example at the top of this page.)

Total Costs Including O&P are developed as follows:
Mat. is **Bare Material Cost** + 10% = **$108** + **$10.80** = **$118.80**
Labor for Crew F2 = Man-hour Cost **($30.15)** × Man-hour Units **(.800)** = **$24.12**
Equip. for Crew F2 = Equip. Hour Cost **($1.45)** × Man-hour Units **(.800)** = **$1.16**
Total = **Mat. Cost ($118.80)** + **Labor Cost ($24.12)** + **Equip. Cost $1.16** = **$144.08**, rounded to **$145.00** per door.
(**Note:** Equipment and Labor costs are derived from the Crew Tables. See example at top of this page. Total line follows the rounding rules.)

8.1 Metal Doors

			DAILY OUTPUT	MAN-HOURS	UNIT	BARE COSTS				TOTAL INCL O&P	
		CREW				MAT.	LABOR	EQUIP.	TOTAL		
100	5840	8'-0" wide, double	F-2	12	1.330	Ea.	94	27	1.77	122.77	145	100
	6200	8-¾" deep, 7'-0" high, 4'-0" wide, single		15	1.070		87	22	1.41	110.41	130	
	6240	8'-0" wide, double	↓	12	1.330		105	27	1.77	133.77	160	
	6300	For "A" label use same price as "B" label										
	6400	For baked enamel finish, add					30%	90%				
	6500	For galvanizing, add					10%					
	6600	For porcelain enamel finish, add					100%	150%				
	7900	Transom lite frames, fixed, add	F-2	155	.103	S.F.	8	2.12	.14	10.26	12.05	
	8000	Movable, add	*	130	.123	*	9.50	2.53	.16	12.19	14.35	
210	0010	COMMERCIAL STEEL DOORS Flush, full panel										210
	0020	Hollow core, 1-¾" thick, 20 ga., 2'-0" x 6'-8"	F-2	20	.800	Ea.	108	16.45	1.06	125.51	145	
	0040	2'-6" x 6'-8"		18	.889		114	18.25	1.18	133.43	155	
	0060	3'-0" x 6'-8"		17	.941		120	19.35	1.25	140.60	160	
	0100	3'-0" x 7'-0"		17	.941		130	19.35	1.25	150.60	175	
	0120	For vision lite, add					50			50	55	
	0140	For narrow lite, add					60			60	66	
	0160	For bottom louver, add				↓	115			115	125	
	0230	For baked enamel finish, add					40%	90%				
	0260	For galvanizing, add					15%					
	0290	For porcelain enamel finish, add					100%	150%				
	0000											
	0320	Half glass, 20 ga., 2'-0" x 6'-8"	F-2	20	.800	Ea.	145	16.45	1.06	162.51	185	
	0340	2'-6" x 6'-8"		18	.889		150	18.25	1.18	169.43	195	
	0360	3'-0" x 6'-8"		17	.941		160	19.35	1.25	180.60	205	
	0400	3'-0" x 7'-0"		17	.941		165	19.35	1.25	185.60	210	
	1020	Hollow core, 1-¾" thick, full panel, 20 ga., 2'-0" x 6'-8"		18	.589		120	18.25	1.18	139.43	160	
	1040	3'-0" x 6'-8"		17	.941		125	19.35	1.25	145.60	165	
	1050	3'-0" x 7'-0"		17	.941		135	19.35	1.25	155.60	180	
	1080	4'-0" x 7'-0"		15	1.070		175	22	1.41	198.41	225	
	1100	4'-0" x 8'-0"		13	1.230		220	25	1.63	246.63	280	
	1120	18 ga., 2'-6" x 6'-8"		17	1		140	19.35	1.25	160.60	185	
	1140	3'-0" x 6'-8"		16	1		150	21	1.33	172.33	195	
	1160	3'-0" x 7'-0"		16	1		160	21	1.33	182.33	210	
	1180	4'-0" x 7'-0"		14	1.140		195	23	1.51	219.51	250	
	1200	4'-0" x 8'-0"		14	1.140		225	23	1.51	249.51	285	
	1220	Half glass, 20 ga., 2'-6" x 6'-8"		20	.800		155	16.45	1.06	172.51	195	
	1240	3'-0" x 6'-8"		18	.889		165	18.25	1.18	184.43	210	
	1260	3'-0" x 7'-0"		18	.889		175	18.25	1.18	194.43	220	
	1260	4'-0" x 7'-0"		16	1		215	21	1.33	237.33	270	

Figure 24.3

iv

218

Installing Contractor's Overhead & Profit

Below are the **average** installing contractor's percentage mark-ups applied to base labor rates to arrive at typical billing rates.

Column A: Labor rates are based on union wages averaged for 30 major U.S. cities. Base rates including fringe benefits are listed hourly and daily. These figures are the sum of the wage rate, employer-paid fringe benefits such as vacation pay, employer-paid health and welfare costs, pension costs, plus appropriate training and industry advancement funds costs.

Column B: Workers' Compensation rates are the national average of state rates established for each trade.

Column C: Column C lists average fixed overhead figures for all trades. Included are Federal and State Unemployment costs set at 5.5%; Social Security Taxes (FICA) set at 7.15%; Builder's Risk Insurance costs set at 1.14%; and Public Liability costs set at 0.82%. All the percentages except those for Social Security Taxes vary from state to state as well as from company to company.

Column D and E: Percentages in Columns D and E are based on the presumption that the installing contractor has annual billing of $500,000 and up. Overhead percentages may increase with smaller annual billing. The overhead percentages for any given contractor may vary greatly and depend on a number of factors, such as the contractor's annual volume, engineering and logistical support costs, and staff requirements. The figures for overhead and profit will also vary depending on the type of job, the job location, and the prevailing economic conditions. All factors should be examined very carefully for each job.

Column F: Column F lists the total of columns B, C, D, and E.

Column G: Column G is Column A (hourly base labor rate) multiplied by the percentage in Column F (O&P percentage).

Column H: Column H is the total of Column A (hourly base labor rate) plus Column G (Total O&P).

Column I: Column I is Column H multiplied by eight hours.

		A		B	C	D	E	F		G	H		I
		Base Rate Incl. Fringes		Work-ers' Comp. Ins.	Average Fixed Over-head	Over-head	Profit	Total Overhead & Profit			Rate with O & P		
Abbr.	Trade	Hourly	Daily					%	Amount		Hourly	Daily	
Skwk	Skilled Workers Average (35 trades)	$20.80	$166.40	10.4%	14.6%	12.6%	10%	47.6%	$ 9.90		$30.70	$245.60	
	Helpers Average (5 trades)	15.90	127.20	11.1		12.7		48.4	7.70		23.60	188.80	
	Foremen Average, Inside (50¢ over trade)	21.30	170.40	10.4		12.6		47.6	10.15		31.45	251.60	
	Foremen Average, Outside ($2.00 over trade)	22.80	182.40	10.4		12.6		47.6	10.85		33.65	269.20	
Clab	Common Building Laborers	16.10	128.80	11.4		10.8		46.8	7.55		23.65	189.20	
Asbe	Asbestos Workers	23.00	184.00	8.8		15.7		49.1	11.30		34.30	274.40	
Boil	Boilermakers	23.00	184.00	7.2		16.0		47.8	11.00		34.00	272.00	
Bric	Bricklayers	20.55	164.40	8.5		10.7		43.8	9.00		29.55	236.40	
Brhe	Bricklayer Helpers	16.15	128.40	8.5		10.7		43.8	7.05		23.20	185.60	
Carp	Carpenters	20.55	164.40	11.4		10.8		46.8	9.60		30.15	241.20	
Cefi	Cement Finishers	19.70	157.60	6.6		10.8		42.0	8.30		27.95	223.60	
Elec	Electricians	22.65	181.20	4.5		15.9		45.0	10.20		32.85	262.80	
Elev	Elevator Constructors	23.05	184.40	6.0		15.9		46.5	10.70		33.75	270.00	
Eqhv	Equipment Operators, Crane or Shovel	21.20	169.60	7.8		13.8		46.2	9.80		31.00	248.00	
Eqmd	Equipment Operators, Medium Equipment	20.75	166.00	7.8		13.8		46.2	9.60		30.35	242.80	
Eqlt	Equipment Operators, Light Equipment	19.60	156.80	7.8		13.8		46.2	9.05		28.65	229.20	
Eqol	Equipment Operators, Oilers	17.55	140.40	7.8		13.8		46.2	8.10		25.65	205.20	
Eqmm	Equipment Operators, Master Mechanics	22.00	176.00	7.8		13.8		46.2	10.15		32.15	257.20	
Glaz	Glaziers	20.75	166.00	8.7		10.8		44.1	9.15		29.90	239.20	
Lath	Lathers	20.50	164.00	7.2		10.8		42.6	8.75		29.25	234.00	
Marb	Marble Setters	20.25	162.00	8.5		10.7		43.8	8.85		29.10	232.80	
Mill	Millwrights	21.25	170.00	7.2		10.8		42.6	9.05		30.30	242.40	
Mstz	Mosaic and Terrazzo Workers	20.05	160.40	6.0		10.7		41.3	8.30		28.35	226.80	
Pord	Painters, Ordinary	19.55	156.40	8.9		10.8		44.3	8.65		28.20	225.60	
Psst	Painters, Structural Steel	20.20	161.60	29.4		10.1		64.1	12.95		33.15	265.20	
Pape	Paper Hangers	19.75	158.00	8.9		10.8		44.3	8.75		28.50	228.00	
Pile	Pile Drivers	20.35	162.80	18.3		15.7		58.6	11.95		32.30	258.40	
Plas	Plasterers	20.25	162.00	8.7		10.9		44.2	8.95		29.20	233.60	
Plah	Plasterer Helpers	16.65	133.20	8.7		10.9		44.2	7.35		24.00	192.00	
Plum	Plumbers	23.00	184.00	5.3		15.9		45.8	10.55		33.55	268.40	
Rodm	Rodmen (Reinforcing)	22.10	176.80	18.6		13.3		56.5	12.50		34.60	276.80	
Rofc	Roofers, Composition	19.15	153.20	20.8		10.4		55.8	10.70		29.85	238.80	
Rots	Roofers, Tile & Slate	19.25	154.00	20.8		10.4		55.8	10.75		30.00	240.00	
Rohe	Roofer Helpers (Composition)	14.20	113.60	20.8		10.4		55.8	7.90		22.10	176.80	
Shee	Sheet Metal Workers	23.10	184.80	7.0		15.8		47.4	10.95		34.05	272.40	
Spri	Sprinkler Installers	23.85	190.80	6.1		15.9		46.6	11.10		34.95	279.60	
Stpi	Steamfitters or Pipefitters	23.30	186.40	5.3		15.9		45.8	10.70		33.95	271.60	
Ston	Stone Masons	20.60	164.80	8.5		10.7		43.8	9.00		29.60	236.80	
Sswk	Structural Steel Workers	22.10	176.80	21.5		13.7		59.8	13.20		35.30	282.40	
Tilf	Tile Layers (Floor)	20.00	160.00	6.0		10.7		41.3	8.25		28.25	226.00	
Tilh	Tile Layer Helpers	16.10	128.80	6.0		10.7		41.3	6.65		22.75	182.00	
Trlt	Truck Drivers, Light	16.70	133.60	9.8		10.7		45.1	7.55		24.25	194.00	
Trhv	Truck Drivers, Heavy	16.95	135.60	9.8		10.7		45.1	7.65		24.60	196.80	
Sswl	Welders, Structural Steel	22.10	176.80	21.5		13.7		59.8	13.20		35.30	282.40	
Wrck	*Wrecking	16.10	128.80	23.1	▼	10.4	▼	58.1	9.35		25.45	203.60	

*Not included in Averages.

Figure 24.4

4. Daily equipment costs, based on the weekly equipment rental cost divided by 5, plus the hourly operating cost, times 8 hours. This cost is listed two ways: as a bare cost and with a 10 percent markup to cover handling and management costs.
5. Labor and equipment are broken down further into cost per man-hour for labor, and cost per man-hour for the equipment.
6. The total daily man-hours for the crew.
7. The total bare costs per day for the crew, including equipment.
8. The total daily cost of the crew including the installing contractor's overhead and profit.

The total daily cost of the required crew is used to calculate the unit installation cost for each item (for both bare costs and cost including overhead and profit).

The crew designation does not mean that this is the only crew that can perform the work. Crew size and content have been developed and chosen based on practical experience and feedback from contractors. These designations represent a labor and equipment make-up commonly found in the industry. The most appropriate crew for a given task is best determined based on particular project requirements. Unit costs may vary if crew sizes or content are significantly changed.

Unit

The unit column (see Figures 24.2 and 24.3) defines the component for which the costs have been calculated. It is this "unit" on which Unit Price Estimating is based. The units as used represent standard estimating and quantity takeoff procedures. However, the estimator should always check to be sure that the units taken off are the same as those priced. A list of standard abbreviations is included at the back of Means *Interior Cost Data.*

Bare Costs

The four columns listed under "Bare Costs" — "Material", "Labor", "Equipment", and "Total", represent the actual cost of construction items to the contractor. In other words, bare costs are those which *do not* include the overhead and profit of the installing contractor, whether for a subcontractor or a general contracting company using its own crews.

Material costs are based on the national average contractor purchase price delivered to the job site. Delivered costs are assumed to be within a 20 mile radius of metropolitan areas. No sales tax is included in the material prices because of variations from state to state.

The prices are based on quantities that would normally be purchased for interior projects costing $20,000 and up. Prices for small quantities must be adjusted accordingly. If more current costs for materials are available for the appropriate location, it is recommended that adjustments be made to the unit costs to reflect any cost difference.

Labor costs are calculated by multiplying the "Bare Labor Cost" per man-hour times the number of man-hours, from the "Man-Hours" column. The "Bare" labor rate is determined by adding the the base rate plus fringe benefits. The base rate is the actual hourly wage of a worker used in figuring payroll. It is from this figure that employee deductions are taken (Federal withholding, FICA, State withholding). Fringe benefits include all employer-paid benefits, above and beyond the payroll amount (employer-paid health, vacation pay, pension, profit-sharing). The "Bare Labor Cost" is, therefore, the actual amount that the contractor must pay directly for construction workers. Figure 24.4 shows labor rates for the 35 standard construction rates plus skilled worker, helper, and foreman averages. These rates are the averages of union wage agreements effective January 1 of the current year from 30 major cities in the United States. The "Bare Labor Cost" for each trade, as used in *Interior Cost Data* is shown in column "A" as the base rate including fringes. Refer to the "Crew" column to determine what rate is used to calculate the "Bare Labor Cost" for a particular line item.

Equipment costs are calculated by multiplying the "Bare Equipment Cost" per man-hour, from the appropriate "Crew" listing, times the man-hours in the "Man-Hours" column.

Total Bare Costs
This column simply represents the arithmetic sum of the bare material, labor, and equipment costs. This total is the average cost to the contractor for the particular item of construction, furnished and installed, or "in place". No overhead and/or profit is included.

Total Including Overhead and Profit
The prices in the "Total Including Overhead and Profit" column represent the total cost of an item including the installing contractor's overhead and profit. The installing contractor could be either the general contractor or a subcontractor. If these costs are used for an item to be installed by a subcontractor, the general contractor should include an additional percentage (usually 10% to 20%) to cover the expenses of supervision and management.

The costs in this column are the arithmetical sum of the following three calculations:

1. Bare Material Cost plus 10%.
2. Labor Cost, including fixed overhead, overhead, and profit, per man-hour times the number of man-hours.
3. Equipment Cost plus 10%, per man-hour times the number of man-hours. The Labor and Equipment Costs, including overhead and profit are found in the appropriate crew listings. The overhead and profit percentage factor for labor is obtained from column F in Figure 24.4.

The following items are included in the increase for fixed overhead, overhead, and profit:

- *Workers' Compensation Insurance* rates vary from state to state and are tied into the construction trade safety records in that particular state. Rates also vary by trade according to the hazard involved. The proper authorities will most likely keep the contractor well informed of the rates and obligations.
- *State and Federal Unemployment Insurance* rates are adjusted by a merit-rating system according to the number of former employees applying for benefits. Contractors who find it possible to offer a maximum of steady employment can enjoy a reduction in the unemployment tax rate.
- *Employer-Paid Social Security* (FICA) is adjusted annually by the federal government. It is a percentage of an employee's salary up to a maximum annual contribution.
- *Builder's Risk and Public Liability* insurance rates vary according to the trades involved and the state in which the work is done.

Overhead is an average percentage to be added for office or operating overhead. This is the cost of doing business. The percentages are presented as national averages by trade as shown in Figure 24.4. Note that the operating overhead costs are applied to *labor only* in Means *Interior Cost Data*.

Profit is the fee (usually a percentage) added by the contractor to offer both a return on investment and an allowance to cover the risk involved in the type of construction being bid. The profit percentage may vary from 4% on large, straightforward projects to as much as 25% on smaller, high-risk jobs. Profit percentages are directly affected by economic conditions, the expected number of bidders, and the estimated risk involved in the project. For estimating purposes, Means *Interior Cost Data* assumes 10% as a reasonable average profit factor.

Section B — Estimating References

Throughout the Unit Price section are circled reference numbers. These numbers serve as footnotes, referring the reader to illustrations, charts, and estimating reference tables in Section B. Figure 24.5 shows an example reference number (for painting) as it appears on a Unit Price page. Figure 24.6 shows the corresponding reference page from Section B. The development of unit costs for many items is explained in these reference tables. Design criteria for many types of interior construction are also included to aid the designer/estimator in making appropriate choices.

Section C — Assemblies Cost Tables

Means' systems data are divided into twelve "Uniformat" divisions, which reorganize the components of construction into logical groupings. The Systems approach was devised to provide quick and easy methods for estimating even when only preliminary design data is available.

The groupings, or systems, are presented in such a way so that the estimator can easily vary components within the systems as well as substituting one system for another. This is extremely useful when adapting to budget, design, or other considerations. Figure 24.7 shows how the data is presented in the Systems section.

Each system is illustrated and accompanied by a detailed description. The book lists the components and sizes of each system, usually in the order of construction. Alternates for the most commonly variable components are also listed. Each individual component is found in the Unit Price Section. If an alternate (not listed in Systems) is required, it can easily be substituted.

Quantity
A unit of measure is established for each system. For example, partition systems are measured by the square foot of wall area, and are measured by "each". Within each system, the components are measured by industry standard, using the same units as in the Unit Price section.

Material
The cost of each component in the Material column is the "Bare Material Cost", plus 10% handling, for the unit and quantity as defined in the "Quantity" column.

Installation
Installation costs as listed in the Systems pages contain both labor and equipment costs. The labor rate includes the "Bare Labor Cost" plus the installing contractor's fixed overhead, overhead, and profit (shown in Figure 24.4). The equipment rate is the "Bare Equipment Cost", plus 10%.

9.8		Painting & Wall Covering	CREW	DAILY OUTPUT	MAN-HOURS	UNIT	MAT.	LABOR	EQUIP.	TOTAL	TOTAL INCL O&P	
050	5000	For latex paint, deduct				S.F.	10%					050
150	0010	CORNER GUARDS Rubber, 3" wide, standard	1 Pord	135	.059	L.F.	2.20	1.16		3.36	4.09	150
	0100	¼" thick, 2-¾" wide		135	.059		2.50	1.16		3.66	4.42	
	0300	Bullnose		135	.059		3.70	1.16		4.86	5.75	
	0400	Vinyl, 5/16" thick, 2-½" wide		135	.059		2.15	1.16		3.31	4.04	
170	0010	DOORS AND WINDOWS										170
	0020	Labor cost includes protection of adjacent items not painted										
	0500	Flush door and frame, per side, oil base, primer coat, brushwork	1 Pord	14	.571	Ea.	1.05	11.15		12.20	17.30	
	1000	Paint, 1 coat		13	.615		1.31	12.05		13.36	18.80	
	1200	2 coats (115)		7.50	1.070		2.55	21		23.55	33	
	1220	3 coats		5.50	1.450		3.70	28		31.70	45	
	1400	Stain, brushwork, wipe off (116)		15	.533		.98	10.45		11.43	16.10	
	1600	Shellac, 1 coat, brushwork		12	.667		.89	13.05		13.94	19.80	
	1800	Varnish, 3 coats, brushwork		5	1.600		2.70	31		33.70	48	
	2000	Panel door and frame, per side, oil base, primer coat, brushwork		10	.800		1.15	15.65		16.80	24	
	2200	Paint, 1 coat		9	.889		1.41	17.40		18.81	27	
	2400	2 coats		5	1.600		2.81	31		33.81	48	
	2420	3 coats		3.50	2.290		4	45		49	69	
	2600	Stain, brushwork, wipeoff		10	.800		1.08	15.65		16.73	24	
	2800	Shellac, 1 coat, brushwork		9	.889		.98	17.40		18.38	26	
	3000	Varnish, 3 coats, brushwork		3.50	2.290		2.95	45		47.95	68	
	4400	Windows, including frame and trim, per side										
	4600	Colonial type, 2' x 3', oil base, primer coat, brushwork	1 Pord	24	.333	Ea.	.85	6.50		7.35	10.35	
	5800	Paint, 1 coat		22	.364		1	7.10		8.10	11.35	
	6000	2 coats		13	.615		1.95	12.05		14	19.50	
	6010	3 coats		10	.800		2.80	15.65		18.45	26	
	6100											
	6200	3' x 5' opening, primer coat, brushwork	1 Pord	15	.533	Ea.	1.05	10.45		11.50	16.20	
	6400	Paint, 1 coat		13	.615		1.26	12.05		13.31	18.75	
	6600	2 coats		8	1		2.43	19.55		21.98	31	
	6610	3 coats		5.50	1.450		3.30	28		31.30	45	
	6800	4' x 8' opening, primer coat, brushwork		12	.667		1.26	13.05		14.31	20	
	7000	Paint, 1 coat		10	.800		1.51	15.65		17.16	24	
	7200	2 coats		6	1.330		2.94	26		28.94	41	
	7210	3 coats		4	2		4.25	39		43.25	61	
	7500	Standard, 6 to 8 lites, 2'x3', primer		26	.308		.80	6		6.80	9.55	
	7520	Paint 1 coat		24	.333		1	6.50		7.50	10.50	
	7540	2 coats		14	.571		1.95	11.15		13.10	18.25	
	7560	3 coats		10	.800		2.70	15.65		18.35	26	
	7580	3' x 5', primer		17	.471		1.02	9.20		10.22	14.40	
	7600	Paint 1 coat		15	.533		1.25	10.45		11.70	16.40	
	7620	2 coats		9	.889		2.43	17.40		19.83	28	
	7640	3 coats		6	1.330		3.60	26		29.60	42	
	7660	4' x 8', primer		14	.571		1.25	11.15		12.40	17.50	
	7680	Paint 1 coat		12	.667		1.51	13.05		14.56	20	
	7700	2 coats		7	1.140		2.94	22		24.94	35	
	7720	3 coats		5	1.600		4.30	31		35.30	50	
	8000	Single lite type, 2' x 3', oil base, primer coat, brushwork		40	.200		.80	3.91		4.71	6.50	
	8200	Paint, 1 coat		37	.216		.97	4.23		5.20	7.15	
	8400	2 coats		21	.381		1.87	7.45		9.32	12.80	
	8410	3 coats		15	.533		2.70	10.45		13.15	18	
	8600	3' x 5' opening, primer coat, brushwork		27	.296		1	5.80		6.80	9.45	
	8800	Paint, 1 coat		25	.320		1.23	6.25		7.48	10.40	
	9000	2 coats		14	.571		2.39	11.15		13.54	18.75	
	9010	3 coats		10	.800		3.69	15.65		19.34	27	
	9200	4' x 8' opening, primer coat, brushwork		17	.471		2.25	9.20		11.45	15.75	
	9400	Paint, 1 coat		15	.533		2.45	10.45		12.90	17.75	
	9600	2 coats		10	.800		3.79	15.65		19.44	27	
	9610	3 coats		8	1		5.10	19.55		24.65	34	

118

Figure 24.5

CIRCLE REFERENCE NUMBERS

(115) Paint (Div. 9.8)

Material prices per gallon in 5 gallon lots, up to 25 gallons. For 100 gallons, deduct 10%.

Exterior, Alkyd (oil base)	
Flat	$14.15
Gloss	16.50
Primer	15.95

Exterior, Latex (water base)	
Acrylic stain	10.95
Gloss enamel	19.00
Flat	13.95
Primer	14.95
Semi-gloss	15.75

Interior, Alkyd (oil base)	
Enamel undercoater	13.55
Flat	12.85
Gloss	18.20
Primer sealer	11.95
Semi-gloss	15.25

Interior, Latex (water base)	
Enamel undercoater	10.80
Flat	10.45
Floor and deck	14.45
Gloss	15.75
Primer sealer	8.65
Semi-gloss	14.30

Masonry, Exterior	
Alkali resistant primer	$13.95
Block filler, epoxy	16.75
Block filler, latex	8.25
Latex, flat or semi-gloss	13.95

Masonry, Interior	
Alkali resistant primer	13.95
Block filler, epoxy	16.75
Block filler, latex	8.25
Floor, alkyd	12.65
Floor, latex	13.75
Latex, flat acrylic	8.55
Latex, flat emulsion	10.45
Latex, sealer	8.65
Latex, semi-gloss	14.30

Varnish and Stain	
Alkyd clear	14.35
Polyurethane, clear	14.75
Primer sealer	11.75
Semi-transparent stain	11.00
Solid color stain	11.95

Metal Coatings	
Galvanized	17.95
High heat	30.00
Machinery enamel, alkyd	18.50
Normal heat	13.25

Rust inhibitor ferrous metal	$16.75
Zinc chromate	17.95

Heavy Duty Coatings	
Acrylic urethane	46.50
Chlorinated rubber	23.00
Coal tar epoxy	16.50
Metal pretreatment (polyvinyl butyral)	19.25
Polyamide epoxy finish	25.95
Polyamide epoxy primer	29.95
Silicone alkyd	30.00
2 component solvent based acrylic epoxy	26.75
2 component solvent based polyester epoxy	23.00
Vinyl	19.25
Zinc rich primer	35.85

Special Coatings/Miscellaneous	
Aluminum	14.60
Creosote	7.00
Dry fall out, flat	8.50
Fire retardant, intumescent	22.00
Linseed oil	8.25
Shellac	12.50
Swimming pool, epoxy or urethane base	30.00
Swimming pool, rubber base	20.75
Texture paint	8.85
Turpentine	8.80
Water repellent 5% silicone	9.60

(116) Painting (Div. 9.8)

Item	Coat	One Gallon Covers			In 8 Hrs. Man Covers			Man Hours per 100 S.F.		
		Brush	Roller	Spray	Brush	Roller	Spray	Brush	Roller	Spray
Paint wood siding	prime	275 S.F.	250 S.F.	325 S.F.	1150 S.F.	1400 S.F.	4000 S.F.	.695	.571	.200
	others	300	275	325	1600	2200	4000	.500	.364	.200
Paint exterior trim	prime	450	—	—	650	—	—	1.230	—	—
	1st	525	—	—	700	—	—	1.143	—	—
	2nd	575	—	—	750	—	—	1.067	—	—
Paint shingle siding	prime	300	285	335	1050	1700	2800	.763	.470	.286
	others	400	375	425	1200	2000	3200	.667	.400	.250
Stain shingle siding	1st	200	190	220	1200	1400	3200	.667	.571	.250
	2nd	300	275	325	1300	1700	4000	.615	.471	.200
Paint brick masonry	prime	200	150	175	850	1700	4000	.941	.471	.200
	1st	300	250	320	1200	2200	4400	.364	.364	.182
	2nd	375	340	400	1300	2400	4400	.615	.333	.182
Paint interior plaster or drywall	prime	450	425	550	1600	2500	4000	.500	.320	.200
	others	500	475	550	1400	3000	4000	.571	.267	.200
Paint interior doors and windows	prime	450	—	—	1300	—	—	.333	—	—
	1st	475	—	—	1150	—	—	.696	—	—
	2nd	500	—	—	1000	—	—	.800	—	—

Figure 24.6

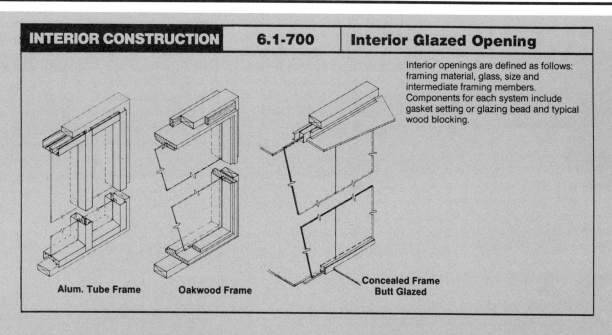

Interior openings are defined as follows: framing material, glass, size and intermediate framing members. Components for each system include gasket setting or glazing bead and typical wood blocking.

Alum. Tube Frame   **Oakwood Frame**   **Concealed Frame Butt Glazed**

System Components	QUANTITY	UNIT	COST PER OPNG.		
			MAT.	INST.	TOTAL
SYSTEM 06.1-700-1000					
GLAZED OPENING, ALUMINUM TUBE FRAME, FLUSH, ¼" FLOAT GLASS, 6' X 4'					
Aluminum tube frame, flush, anodized bronze, head & jamb	16.500	L.F.	76.23	98.67	174.90
Aluminum tube frame, flush, anodized bronze, open sill	7.000	L.F.	32.76	40.74	73.50
Joints for tube frame, clip type	4.000	Ea.	15.40		15.40
Gasket setting, add	20.000	L.F.	43		43
Wood blocking	8.000	B.F.	2.95	11.25	14.20
Float glass, ¼" thick, clear, plain	24.000	S.F.	42.24	95.76	138
TOTAL			212.58	246.42	459

6.1-700		Interior Glazed Opening						
	FRAME	GLASS	OPENING-SIZE W X H	INTERMEDIATE MULLION	INTERMEDIATE HORIZONTAL	COST PER OPNG.		
						MAT.	INST.	TOTAL
1000	Aluminum flush	¼" float	6'x4'	0	0	215	245	460
1040	Tube		12'x4'	3	0	515	530	1,045
1080			4'x5'	0	0	190	215	405
1120			8'x5'	1	0	355	395	750
1160			12'x5'	2	0	520	575	1,095
1240		⅜" float	9'x6'	2	0	630	660	1,290
1280			4'x8'-6"	0	1	405	425	830
1320			16'x10'	3	1	1,675	1,750	3,425
1400		¼" tempered	6'x4'	0	0	270	245	515
1440			12'x4'	3	0	630	525	1,155
1480			4'x5'	0	0	240	210	450
1520			8'x5'	1	0	450	395	845
1560			12'x5'	2	0	665	575	1,240
1640		⅜" tempered	9'x6'	2	0	995	660	1,655
1680			4'x8'-6"	0	0	580	400	980
1720			16'x10'	3	0	2,550	1,625	4,175
1800		¼" one way mirror	6'x4'	0	0	460	245	705
1840			12'x4'	3	0	1,000	520	1,520
1880			4'x5'	0	0	395	210	605
1920			8'x5'	1	0	770	390	1,160

296

Figure 24.7

City Cost Indexes

The unit prices in Means *Interior Cost Data* are national averages. When they are to be applied to a particular location, these prices must be adjusted to local conditions. R. S. Means has developed the City Cost Indexes for just that purpose. Section D of Means *Interior Cost Data* contains tables of indexes for 162 U.S. and Canadian cities based on a 30 major city average of 100. The figures are broken down into material and installation for all CSI Divisions, as shown in Figure 24.8. Please note that for each city there is a weighted average based on total project costs. This average is based on the relative contribution of each division to the construction process as a whole.

In addition to adjusting the figures in Means *Interior Cost Data* for particular locations, the City Cost Index can also be used to adjust costs from one city to another. For example, interior costs for a particular building type is known for City A. In order to budget the costs of the same building type in City B, the following calculation can be made:

$$\frac{\text{City B Index}}{\text{City A Index}} \quad x \quad \text{City A Cost} \quad = \quad \text{City B Cost}$$

While City Cost Indexes provide a means to adjust prices for location, the Historical Cost Index, (also included in Means *Interior Cost Data* and shown in Figure 24.9) provides a means to adjust for time. Using the same principle as above, a time-adjustment factor can be calculated:

$$\frac{\text{Index for Year X}}{\text{Index for Year Y}} \quad = \quad \text{Time-adjustment Factor}$$

This time-adjustment factor can be used to determine the budget costs for a particular building type in Year X, based on costs for a similar building type known from Year Y. Used together, the two indexes allow for cost adjustments from one city during a given year to another city in another year (the present or otherwise). For example, an office building built in San Francisco in 1974, originally cost $1,000,000. How much will a similar building cost in Phoenix in 1987? Adjustment factors are developed as shown above using data from Figures 24.8 and 24.9:

$$\frac{\text{Phoenix index}}{\text{San Francisco index}} \quad = \quad \frac{92.8}{123.4} \quad = \quad 0.75$$

$$\frac{\text{1987 index}}{\text{1974 index}} \quad = \quad \frac{195.8}{94.7} \quad = \quad 2.07$$

Original cost x location adjustment x time adjustment = Proposed new cost

$$\$1,000,000 \quad x \quad 0.75 \quad x \quad 2.07 \quad = \quad \$1,552,500$$

DIVISION		ALABAMA										ALASKA			ARIZONA		
		BIRMINGHAM			HUNTSVILLE			MOBILE			MONTGOMERY			ANCHORAGE			PHOENIX	
		MAT.	INST.	TOTAL	MAT.	INST.	TOTAL	MAT.	INST.	TOTAL	MAT.	INST.	TOTAL	MAT.	INST.	TOTAL	MAT.	INST.	TOTAL
2	SITE WORK	96.7	91.2	94.3	115.6	92.1	105.1	118.5	88.9	105.3	88.1	87.7	87.9	154.7	131.7	144.4	89.9	95.0	92.2
3.1	FORMWORK	90.5	72.6	76.5	92.7	73.4	77.6	97.0	78.7	82.7	102.1	70.2	77.2	114.1	146.3	139.3	108.4	91.8	95.4
3.2	REINFORCING	94.6	78.8	88.0	95.8	73.3	86.4	83.0	77.9	80.9	83.0	78.8	81.2	117.8	140.5	127.3	111.7	105.7	109.2
3.3	CAST IN PLACE CONC.	89.4	92.0	91.0	102.0	93.6	96.8	100.1	94.6	96.7	101.4	90.8	94.8	226.0	115.0	157.7	107.7	93.4	98.9
3	CONCRETE	90.7	83.2	85.9	98.8	83.9	89.2	95.7	86.9	90.1	97.5	81.7	87.3	180.4	129.5	147.7	108.7	93.9	99.2
4	MASONRY	79.9	76.7	77.5	86.1	69.8	73.7	92.2	84.5	86.3	85.2	56.9	63.6	137.0	149.3	146.4	93.3	69.5	75.1
5	METALS	95.8	84.0	91.6	100.4	80.1	93.1	93.7	84.2	90.3	96.1	83.9	91.7	116.7	130.9	121.7	98.7	101.1	99.6
6	WOOD & PLASTICS	91.7	73.7	81.5	104.1	73.6	86.8	92.0	80.4	85.5	101.4	73.9	85.9	117.9	141.2	131.1	99.6	90.7	94.6
7	MOISTURE PROTECTION	84.6	68.4	79.5	92.0	69.7	85.0	87.1	71.7	82.3	88.3	68.2	82.0	102.5	143.5	115.3	92.7	89.6	91.7
8	DOORS, WINDOWS, GLASS	91.0	74.3	82.3	101.5	67.4	83.8	99.0	78.8	88.5	98.3	72.1	84.7	129.4	136.8	133.3	103.8	89.3	96.3
9.1	LATH & PLASTER	96.1	71.2	77.2	87.0	73.3	76.6	91.9	86.0	87.5	108.4	73.7	82.1	120.4	148.3	141.6	93.1	93.5	93.4
9.2	DRYWALL	100.5	73.4	87.7	108.5	71.9	91.2	92.5	81.5	87.3	100.7	74.5	88.3	121.9	143.8	132.3	90.8	90.1	90.5
9.5	ACOUSTICAL WORK	98.7	73.3	84.9	101.1	73.2	86.0	94.2	80.0	86.5	94.2	73.0	82.7	125.6	142.7	134.9	103.1	89.8	95.9
9.6	FLOORING	112.0	79.6	103.5	97.4	70.7	90.4	114.0	86.4	106.8	100.4	50.7	87.4	117.3	149.3	125.6	93.0	74.5	88.2
9.8	PAINTING	104.5	71.9	78.4	110.7	70.5	78.5	121.5	81.6	89.5	119.7	80.2	88.0	123.2	146.3	141.7	96.5	91.2	92.3
9	FINISHES	103.3	73.2	87.1	105.1	71.5	87.0	100.4	82.0	90.5	102.2	74.7	87.4	121.3	145.3	134.2	92.9	89.6	91.1
10-14	TOTAL DIV. 10-14	100.0	75.4	92.6	100.0	73.2	91.9	100.0	82.7	94.8	100.0	71.4	91.4	100.0	143.2	112.9	100.0	89.8	96.9
15	MECHANICAL	96.5	77.7	86.9	99.6	77.4	88.3	97.5	79.8	88.4	99.2	74.6	86.6	107.5	136.6	122.4	98.5	85.4	91.8
16	ELECTRICAL	94.1	78.1	82.9	92.1	77.7	82.0	90.0	82.1	84.5	90.9	66.4	73.7	108.4	147.0	135.5	105.0	88.9	93.7
1-16	WEIGHTED AVERAGE	94.4	78.6	85.8	99.6	76.9	87.3	97.3	82.9	89.5	96.3	73.1	83.7	124.8	139.2	132.6	99.1	87.5	92.8

DIVISION		ARIZONA			ARKANSAS						CALIFORNIA								
		TUCSON			FORT SMITH			LITTLE ROCK			ANAHEIM			BAKERSFIELD			FRESNO		
		MAT.	INST.	TOTAL	MAT.	INST.	TOTAL	MAT.	INST.	TOTAL	MAT.	INST.	TOTAL	MAT.	INST.	TOTAL	MAT.	INST.	TOTAL
2	SITE WORK	106.6	97.0	102.3	96.6	91.8	94.4	103.3	94.1	99.2	101.0	113.1	106.4	93.1	111.2	101.2	91.7	120.5	104.5
3.1	FORMWORK	100.7	89.7	92.1	102.5	65.7	73.7	95.7	70.9	76.3	94.9	129.7	122.1	112.8	129.7	126.0	99.6	123.5	118.3
3.2	REINFORCING	95.1	105.7	99.5	124.6	71.3	102.3	117.8	69.1	97.5	99.3	130.1	112.2	96.1	130.1	110.3	106.5	130.1	116.3
3.3	CAST IN PLACE CONC.	105.7	97.7	100.7	90.6	91.9	91.4	98.5	92.3	94.7	109.5	109.2	109.3	103.4	109.4	107.1	93.1	107.3	101.9
3	CONCRETE	102.4	95.2	97.8	100.4	79.8	87.2	102.2	81.9	89.2	104.4	119.1	113.8	103.6	119.2	113.8	97.3	115.7	109.1
4	MASONRY	90.2	69.4	74.3	93.5	78.4	81.9	89.8	78.8	81.4	106.6	127.3	122.4	98.8	115.8	111.8	114.7	111.9	112.6
5	METALS	91.1	102.7	95.3	96.9	79.0	90.5	106.6	77.7	96.3	99.5	122.1	107.6	99.7	122.4	107.8	95.3	123.5	105.3
6	WOOD & PLASTICS	107.5	87.9	96.4	106.9	65.8	83.6	94.4	72.2	81.9	95.9	126.5	113.2	95.2	126.5	112.9	96.9	120.1	110.0
7	MOISTURE PROTECTION	105.6	77.5	96.8	84.7	67.3	79.2	84.2	68.3	79.2	108.2	131.6	115.5	84.7	117.6	95.0	107.7	108.2	107.8
8	DOORS, WINDOWS, GLASS	88.4	88.3	88.4	93.2	61.6	76.8	95.7	65.3	79.9	93.8	127.3	111.2	100.4	123.4	112.4	101.4	120.6	111.4
9.1	LATH & PLASTER	109.1	93.8	97.5	93.0	76.0	80.1	98.5	78.4	83.3	97.1	132.4	123.8	95.4	110.3	106.7	101.9	126.3	120.4
9.2	DRYWALL	82.1	88.7	85.2	95.1	64.6	80.7	114.8	71.3	94.2	97.4	127.9	111.9	98.0	121.7	109.2	98.7	124.0	110.7
9.5	ACOUSTICAL WORK	114.9	87.4	99.9	84.5	64.5	73.6	84.5	71.2	77.3	82.2	127.5	106.8	94.1	127.5	112.3	97.4	121.1	110.3
9.6	FLOORING	109.9	74.0	100.5	89.3	78.8	86.6	88.5	80.1	86.3	117.3	126.5	119.7	111.9	112.5	112.1	88.5	106.3	93.2
9.8	PAINTING	98.6	84.1	87.0	111.1	52.9	64.4	104.7	66.8	74.3	108.3	123.9	120.8	120.2	108.1	110.5	108.0	121.8	119.0
9	FINISHES	93.3	86.3	89.5	94.5	62.2	77.1	105.0	70.8	86.5	101.8	126.6	115.2	103.0	116.1	110.1	97.3	121.9	110.6
10-14	TOTAL DIV. 10-14	100.0	87.4	96.2	100.0	74.6	92.4	100.0	75.4	92.6	100.0	128.4	108.5	100.0	125.2	107.5	100.0	146.2	113.8
15	MECHANICAL	98.9	85.4	92.0	97.5	69.3	83.1	97.1	73.6	85.1	96.9	126.2	111.9	95.1	107.5	101.4	92.7	120.9	107.1
16	ELECTRICAL	102.3	87.9	92.2	99.3	76.6	83.3	93.6	80.2	84.2	98.7	139.7	127.5	106.0	103.4	104.2	109.5	101.0	103.5
1-16	WEIGHTED AVERAGE	98.5	86.9	92.2	96.7	74.2	84.5	98.5	77.2	86.9	100.6	126.1	114.5	98.4	114.7	107.3	98.9	117.0	108.7

DIVISION		CALIFORNIA																
		LOS ANGELES			OXNARD			RIVERSIDE			SACRAMENTO			SAN DIEGO			SAN FRANCISCO		
		MAT.	INST.	TOTAL	MAT.	INST.	TOTAL	MAT.	INST.	TOTAL	MAT.	INST.	TOTAL	MAT.	INST.	TOTAL	MAT.	INST.	TOTAL
2	SITE WORK	96.1	115.7	104.9	97.9	106.2	101.6	95.1	111.7	102.5	82.9	105.4	93.0	94.9	108.3	100.9	102.6	114.6	108.0
3.1	FORMWORK	112.9	130.1	126.4	90.1	130.0	121.3	102.5	129.7	123.8	101.0	123.7	118.7	105.2	127.5	122.6	104.9	133.7	127.4
3.2	REINFORCING	64.4	130.1	91.8	99.3	130.1	112.2	124.6	130.1	126.9	99.3	130.1	112.2	119.3	130.1	123.8	124.1	130.1	126.6
3.3	CAST IN PLACE CONC.	97.9	112.3	106.8	102.5	109.9	107.1	102.5	109.6	106.9	116.1	106.5	110.2	100.4	105.3	103.4	105.6	116.3	112.1
3	CONCRETE	93.5	120.8	111.1	99.4	119.6	112.4	107.3	119.3	115.0	109.5	115.3	113.2	105.5	116.1	112.3	109.5	124.3	119.0
4	MASONRY	108.8	127.3	122.9	98.4	120.9	115.7	102.8	121.1	116.8	101.4	105.6	104.6	111.3	118.4	116.7	128.6	143.4[1]	139.9
5	METALS	101.8	123.1	109.4	105.6	122.4	111.6	99.6	122.2	107.7	111.4	123.4	115.7	99.0	121.1	106.9	104.2	126.5	112.2
6	WOOD & PLASTICS	100.6	127.6	115.9	92.7	127.1	112.2	94.8	126.5	112.8	78.4	120.4	102.2	97.1	123.8	112.2	94.0	132.6	115.9
7	MOISTURE PROTECTION	103.9	133.5	113.2	89.8	132.4	103.1	90.3	130.7	102.9	85.1	122.6	96.8	94.4	120.1	104.0	100.5	128.7	109.3
8	DOORS, WINDOWS, GLASS	102.2	127.3	115.2	103.0	127.3	115.6	103.6	127.3	115.9	92.2	120.1	106.7	108.1	123.6	116.1	113.4	132.3	123.2
9.1	LATH & PLASTER	95.8	132.5	123.6	97.6	134.0	125.1	97.6	128.3	120.9	98.9	120.1	115.0	102.2	119.7	115.5	101.5	147.1	136.0
9.2	DRYWALL	89.1	127.9	107.5	98.7	130.0	113.5	94.7	127.9	110.5	97.3	120.5	108.3	101.4	124.4	112.3	81.1	136.6	107.4
9.5	ACOUSTICAL WORK	98.7	127.5	114.4	88.4	127.5	109.7	88.4	127.5	109.7	86.6	121.1	105.4	100.7	124.8	113.8	100.7	134.2	118.9
9.6	FLOORING	96.3	126.5	104.2	95.6	126.5	103.7	95.6	126.5	103.7	85.8	133.3	96.3	98.5	121.7	104.6	107.4	140.0	115.9
9.8	PAINTING	84.3	126.3	118.0	92.2	117.8	112.8	100.8	123.9	119.3	112.3	132.4	128.4	92.1	127.9	120.8	102.8	143.5	135.5
9	FINISHES	91.2	127.5	110.7	96.5	125.5	112.1	95.1	126.4	112.0	95.4	125.6	111.6	99.8	125.2	113.5	91.3	139.7	117.4
10-14	TOTAL DIV. 10-14	100.0	128.9	108.6	100.0	128.3	108.5	100.0	128.3	108.5	100.0	147.9	114.3	100.0	126.8	108.0	100.0	153.0	115.9
15	MECHANICAL	97.7	128.8	113.6	98.8	127.4	113.4	96.6	130.9	114.2	98.2	124.1	111.5	103.1	127.6	115.6	100.5	158.4	130.2
16	ELECTRICAL	101.8	132.5	123.4	98.7	157.9	140.3	98.2	134.9	124.0	109.5	99.1	102.2	105.8	108.9	108.0	107.9	151.9	138.8
1-16	WEIGHTED AVERAGE	98.7	126.4	113.7	98.8	127.4	114.3	99.0	125.4	113.3	99.2	116.5	108.6	101.9	119.6	111.5	103.9	139.9	123.4

371

Figure 24.8

19.1 CITY COST INDEXES

Historical Cost Indexes (Div. 1.1-160)

The table below lists both the Means City Cost Index based on Jan. 1, 1975 = 100 as well as the computed value of an index based on January 1, 1987 costs. Since the Jan. 1, 1987 figure is estimated, space is left to write in the actual index figures as they become available thru either the quarterly "Means Construction Cost Indexes" or as printed in the "Engineering News-Record". To compute the actual index based on Jan. 1, 1987 = 100, divide the Quarterly City Cost Index for a particular year by the actual Jan. 1, 1987 Quarterly City Cost Index. Space has been left to advance the index figures as the year progresses.

Year	"Quarterly City Cost Index" Jan. 1, 1975 = 100		Current Index Based on Jan. 1, 1987 = 100		Year	"Quarterly City Cost Index" Jan. 1, 1975 = 100	Current Index Based on Jan. 1, 1987 = 100		Year	"Quarterly City Cost Index" Jan. 1, 1975 = 100	Current Index Based on Jan. 1, 1987 = 100	
	Est.	Actual	Est.	Actual		Actual	Est.	Actual		Actual	Est.	Actual
Oct. 1987					July 1974	94.7	48.4		July 1958	43.0	22.0	
July 1987					1973	86.3	44.1		1957	42.2	21.6	
April 1987					1972	79.7	40.7		1956	40.4	20.6	
Jan. 1987	195.8		100.0	100.0	1971	73.5	37.5		1955	38.1	19.5	
July 1986		192.8	98.5		1970	65.8	33.6		1954	36.7	18.7	
1985		189.1	96.6		1969	61.6	31.5		1953	36.2	18.5	
1984		187.6	95.8		1968	56.9	29.1		1952	35.3	18.0	
1983		183.5	93.7		1967	53.9	27.5		1951	34.4	17.6	
1982		174.3	89.0		1966	51.9	26.5		1950	31.4	16.0	
1981		160.2	81.8		1965	49.7	25.4		1949	30.4	15.5	
1980		144.0	73.5		1964	48.6	24.8		1948	30.4	15.5	
1979		132.3	67.6		1963	47.3	24.2		1947	27.6	14.1	
1978		122.4	62.5		1962	46.2	23.6		1946	23.2	11.8	
1977		113.3	57.9		1961	45.4	23.2		1945	20.2	10.3	
1976		107.3	54.8		1960	45.0	23.0		1944	19.3	9.9	
1975		102.6	52.4		1959	44.2	22.6		1943	18.6	9.5	

City Cost Indexes (Div. 1.1-060)

Tabulated on the following pages are average construction cost indexes for 162 major U.S. and Canadian cities. Index figures for both material and installation are based on the 30 major city average of 100 and represent the cost relationship as of April 1, 1986. The index for each division is computed from representative material and labor quantities for that division. The weighted average for each city is a weighted total of the components listed above it, but does not include relative productivity between trades or cities.

The material index for the weighted average includes about 100 basic construction materials with appropriate quantities of each material to represent typical "average" building construction projects.

The installation index for the weighted average includes the contribution of about 30 construction trades with their representative man-days in proportion to the material items installed. Also included in the installation costs are the representative equipment costs for those items requiring equipment.

Since each division of the book contains many different items, any particular item multiplied by the particular city index may give incorrect results. However, when all the book costs for a particular division are summarized and then factored, the result should be very close to the actual costs for that particular division for that city.

If a project has a preponderance of materials from any particular division (say structural steel), then the weighted average index should be adjusted in proportion to the value of the factor for that division.

370

Figure 24.9

Chapter 25

UNIT PRICE ESTIMATING EXAMPLE

As discussed in the introduction to this book, Unit Price Estimates are the most accurate type of estimate. Final plans and specifications are required to complete the takeoff and pricing. Current costs can be used for every item specified, which accounts for fewer variables and hence, greater accuracy.

The estimator should have all contract documents, including the drawings, specifications, and any addenda, prior to beginning the estimate and should follow procedures recommended in Chapter 17, "Before Starting the Estimate". The design professional or owner should be notified of any discrepancies between the drawings and the specifications; any grey areas should be cleared up at this time. Substitute items should be approved by the design professional or owner, to ensure acceptance when the bid is submitted. These preliminary steps can prevent problems which may otherwise arise before and during construction, and should not be overlooked.

The following example is a Unit Price Estimate for a typical hotel suite. In a case such as this, the estimator can determine quantities and costs for one suite (or for each type of suite) to be multiplied by the total number of suites (or total of each type). Total quantities and costs are thus easily calculated. A description of the project is included below.

All unit costs used in this example are from Means *Interior Cost Data*, 1987. The example assumes that all contract documents are available for developing the estimate and references will be made to them.

Project Description

The sample project shown is a hotel suite renovation. The demolition work has been accomplished and all walls, doors, and cabinets are in place. Interior finish work and furnishings are required and will be estimated. Figures 25.1 and 25.2 show the suite with and without furniture, respectively.

For most projects, the majority of the interior work will be specified in Division 9 – Finishes. Interior lighting (not part of this estimate) would be shown in the reflected ceiling plans and, with other wiring requirements, described in Division 16 of the specifications. Division 12 – Furnishings is usually a separate package from the construction documents and is developed as a separate estimate from the construction portion of the project. In this example the furnishings will be estimated separately and added to the finishes to determine a total cost per suite.

The General Requirements or General Conditions for a project are included as part of the construction contract or in Division I of the specifications. Although written for all trades, the General Requirements must be reviewed to determine all items which are applicable to the interior portion of a project. Such items may include temporary lighting and power, dust protection, temporary partitions, limited access, or special fire protection requirements. Likewise, every other specification section or division should be reviewed for any work that might involve or impact the interior work. The General Requirements can be estimated item by item using unit prices, or costs can be applied as a percentage of the project costs. The latter method, used in this example, should be based on accurate, well developed costs from past, similar projects.

The finishes, even in a small area such as the hotel suite, can involve many different types of materials, treatments, and installation requirements. In order to assure that all items are included, a Room Finish Schedule (shown in Figure 25.3) can be used to organize and list all information.

Floor Finishes

When estimating for flooring, preparation of the subsurface is a cost often overlooked. Floor subsurfaces must be suitable to receive the new finish; some subsurface preparation is usually required. While this is especially important in renovation projects, new floors (concrete or otherwise) must be scraped and cleaned to receive the finish material, which involves additional cost. Surface preparation is most important where resilient floors and thin set ceramic tile floors are installed since even slight uneven subsurfaces are "telegraphed" through the finish material. Carpeted floors may have some tolerance to imperfections, especially where a cushion or pad is installed underneath, but the subsurface must be smooth and hard to ensure even wear.

The estimator should field verify any surface preparation required. Where floor materials change or floor expansion joints exist, a binder bar or transition strip may be required. These items should be included in the estimate.

In this estimating example, wall to wall carpet with resilient base is specified for the living room and bedroom. The bathroom floor is ceramic tile flooring and base, and the kitchen floor is sheet vinyl with resilient base. Figure 25.4 lists the quantities required for each type of flooring specified.

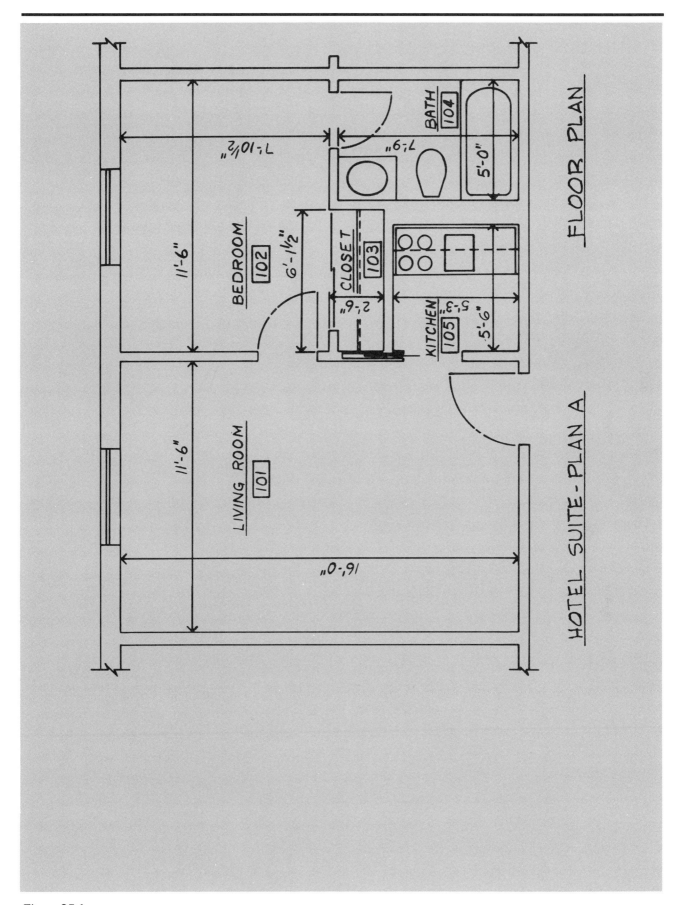

FLOOR PLAN

HOTEL SUITE - PLAN A

Figure 25.1

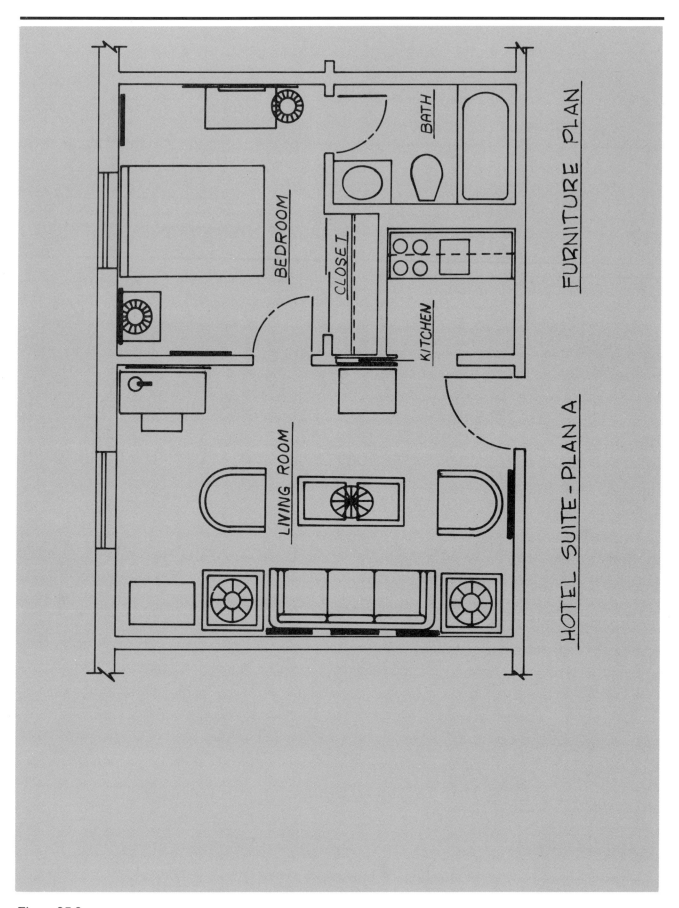

FURNITURE PLAN

HOTEL SUITE - PLAN A

BEDROOM

LIVING ROOM

CLOSET

KITCHEN

BATH

Figure 25.2

232

Means Forms

ROOM FINISH SCHEDULE

PROJECT Hotel Suite - Plan A   ARCHITECT AEL   BY _____   PAGE 1 OF 1   DATE 1987

ROOMS		FLOORS MATERIALS			BASES MATERIALS		WAINSCOTS	WALLS MATERIALS				CEILING MATERIALS		REMARKS
# Room	ROOM NAME	1 Carpet	2 C.T.	3 Vinyl	1 Resiliant	2 C.T.		1 Flat Paint	2 Enamel Pt.	3 V.W.C.	4 Mirrors	1 Textured Spray	2 Enamel Paint	
101	Living Room	X			X			N / S / E		E / N	N	X		Doors & windows Enamel
102	Bedroom	X			X			X				X		
103	Closet	X			X			X				X		
104	Bath		X			X			X				X	
105	Kitchen			X		X			X				X	

Figure 25.3

QUANTITY SHEET

PROJECT **Hotel Suite- Plan A**	SHEET NO. **1 of 1**
	ESTIMATE NO. **87-1**
LOCATION ·	ARCHITECT **AEL** · DATE **1987**
TAKE OFF BY **FWH**	EXTENSIONS BY: **EBW** · CHECKED BY: **SLP**

DESCRIPTION	NO.	DIMENSIONS L	W		Carpet	UNIT	Ceramic	UNIT	Vinyl	UNIT	Base	UNIT
Floor Finishes												
Carpet : (12' wide)												
Living Room		16.0	11.5	12								
Bedroom		8.3	11.5	12								
Closet		2.6	5.8	12								
		27	x	12	324	SF						
5% Contingency					16	SF						
					340	SF						
(note: 2 closets Incl.)					⟨38⟩	SY						
Ceramic :												
Bathroom Floor		5.5	5.0				⟨28⟩	SF				
Bathroom Base		5.5										
		3.0										
		2.5										
		11									⟨11⟩	LF
Resilient :												
Kitchen Floor		5.2	5.5	6					⟨33⟩	SF		
(6' goods)												
Base - 4" Vinyl												
Living Room		47.5										
Bedroom & Closet		30										
Kitchen		18.7										
		96.2										
5%		4.8										
		101									⟨101⟩	LF

Figure 25.4

Carpet

The grade of carpet specified and any special provisions for pattern match should first be determined. Manufacturer, color, type, quality, method of installation, and other special or unique features will be listed in the specifications under Division 9.6 — Flooring, or on the Room Finish Schedule.

A seam diagram should be drawn to determine the amount of carpet needed and acceptable locations for seams. This may be done by using a scale drawing of the room, drawing the carpet seams in acceptable locations. A seam diagram is especially important when the carpet is patterned or is not laid with seams parallel to the wall. The yardage will vary based on the layout. Another factor that may affect the quantity and cost is the manufactured width of the carpet. Note that the width dimension used to calculate carpet quantities is 12', rather than the room dimension of 11'6". Figure 25.5 illustrates the actual pieces of carpet which would be used for the suite. The cross-hatched area in Figure 25.5 includes 5% waste.

The carpet grade specified is 28 oz., nylon, heavy traffic. Since pattern match is not required, only 5% is added to the quantity ordered to allow for waste or any imperfections in room geometry. Costs are entered on the estimate sheet for carpet (Figure 25.6). Note that carpet is taken off and priced by the square yard. A common error such as not converting square feet to square yards can throw the estimate off significantly. Figure 25.6 illustrates the use of a Cost Analysis form. This type of form will be used for the remainder of the estimate.

Ceramic Tile

The specifications for the bathroom floor of the sample hotel suite call for 1" by 1" ceramic tile, color group 2, with 4-1/4" by 4-1/4" cove base. Figure 25.7, from Means *Interior Cost Data*, 1987, shows tile floor costs (see line 9.3-050-3100). Quantities from Figure 25.4 and costs shown in Figure 25.7 are entered on the estimate sheet (Figure 25.6).

Resilient Flooring

The overall room dimensions of the kitchen are calculated for vinyl flooring. Cutouts will be from full sheets with no seams. Quantities must be determined carefully and accurately because the estimate for this suite will be multiplied by the total number of similar suites. Thus any errors will be amplified and could have a significant impact on total costs. As with carpet, the width of the goods must be considered. The quantity and costs for vinyl flooring as well as resilient base are entered on the estimate sheet (Figure 25.6).

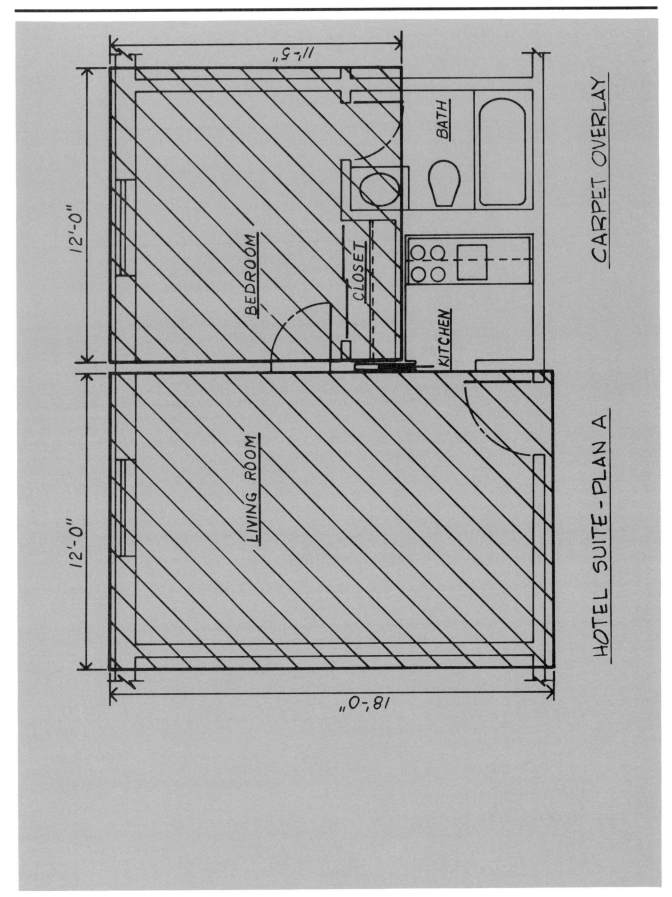

CARPET OVERLAY

HOTEL SUITE – PLAN A

BEDROOM

BATH

CLOSET

KITCHEN

LIVING ROOM

11'-5"

12'-0"

12'-0"

18'-0"

Figure 25.5

Means Forms

COST ANALYSIS

SHEET NO. 1 of 4

PROJECT **Hotel Suite - Plan A**

ESTIMATE NO. **87-1**

ARCHITECT

DATE **1987**

TAKE OFF BY: **FWH** QUANTITIES BY: **FWH** PRICES BY: **RSM** EXTENSIONS BY: **SLP** CHECKED BY: **JDM**

DESCRIPTION	SOURCE/DIMENSIONS			QUANTITY	UNIT	MATERIAL		LABOR		EQ./TOTAL	
						UNIT COST	TOTAL	UNIT COST	TOTAL	UNIT COST	TOTAL
Floor Finishes											
Carpet	9.6	050	3300	38	SY	16^{95}	644	2^{81}	107		
Ceramic - floor	9.3	050	3100	28	SF	1^{78}	50	1^{58}	44		
base	9.3	050	0700	11	LF	1^{45}	16	2^{26}	25		
Sheet Vinyl	9.6	200	8250	33	SF	3^{25}	107	70	23		
4" Vinyl base	9.6	200	1500	101	LF	52	53	51	52		
Total							(870)		(251)		

Figure 25.6

237

9.3 Tile & Terrazzo

		9.3 Tile & Terrazzo	CREW	DAILY OUTPUT	MAN-HOURS	UNIT	BARE COSTS MAT.	BARE COSTS LABOR	BARE COSTS EQUIP.	BARE COSTS TOTAL	TOTAL INCL O&P	
050	0010	**CERAMIC TILE** Base, using 1″ x 1″ tiles, 4″ high, mud set	D-7	82	.195	L.F.	3.20	3.52		6.72	8.50	050
	0100	Thin set	″	128	.125		2.95	2.26		5.21	6.45	
	0300	For 6″ high base, 1″ x 1″ tiles, add					.30			.30	.33	
	0400	For 2″ x 2″ tiles, add to above					.17			.17	.19	
	0600	Cove base, 4-¼″ x 4-¼″ high, mud set	D-7	91	.176		1.55	3.17		4.72	6.20	
	0700	Thin set		128	.125		1.45	2.26		3.71	4.79	
	0900	6″ x 4-¼″ high, mud set		100	.160		1.53	2.89		4.42	5.75	
	1000	Thin set		137	.117		1.39	2.11		3.50	4.51	
	1200	Sanitary cove base, 6″ x 4-¼″ high, mud set		93	.172		1.70	3.11		4.81	6.25	
	1300	Thin set		124	.129		1.65	2.33		3.98	5.10	
	1500	6″ x 6″ high, mud set		84	.190		1.80	3.44		5.24	6.85	
	1600	Thin set		117	.137		1.70	2.47		4.17	5.35	
	2400	Bullnose trim, 4-¼″ x 4-¼″, mud set		82	.195		1.15	3.52		4.67	6.25	
	2500	Thin set		128	.125		1.10	2.26		3.36	4.40	
	2700	6″ x 4-¼″ bullnose trim, mud set		84	.190		1.13	3.44		4.57	6.10	
	2800	Thin set		124	.129		1.10	2.33		3.43	4.50	
	3000	Floors, natural clay, random or uniform, thin set, color group 1		183	.087	S.F.	1.65	1.58		3.23	4.05	
	3100	Color group 2		183	.087		1.78	1.58		3.36	4.19	
	3300	Porcelain type, 1 color, color group 2, 1″ x 1″		183	.087		1.85	1.58		3.43	4.27	
	3400	2″ x 2″ or 2″ x 1″, thin set		190	.084		2.05	1.52		3.57	4.41	
	3600	For random blend, 2 colors, add					.10			.10	.11	
	3700	4 colors, add					.20			.20	.22	
	3900	For color group 3, add					.15			.15	.17	
	4000	For abrasive non-slip tile, add					.42			.42	.46	
	4200	Conductive tile, 1″ squares, black	D-7	109	.147		3.90	2.65		6.55	8.05	
	4220	4″ x 8″ or 4″ x 4″, ⅜″ thick		120	.133		4.10	2.41		6.51	7.90	
	4240	Trim, bullnose, etc.		200	.080	L.F.	1.51	1.44		2.95	3.70	
	4300	Specialty tile, 3″ x 6″ x ½″, decorator finish		183	.087	S.F.	4	1.58		5.58	6.65	
	4500	Add for epoxy grout, 1/16″ joint, 1″ x 1″ tile		800	.020		.61	.36		.97	1.18	
	4600	2″ x 2″ tile		820	.020		.47	.35		.82	1.02	
	4800	Pregrouted sheets, walls, 4-¼″ x 4-¼″, 6″ x 4-¼″										
	4810	and 8-½″ x 4-¼″, 4 S.F. sheets, silicone grout	D-7	240	.067	S.F.	2.10	1.20		3.30	4.01	
	5100	Floors, unglazed, 2 S.F. sheets,										
	5110	urethane adhesive	D-7	180	.089	S.F.	3.35	1.60		4.95	5.95	
	5400	Walls, interior, thin set, 4-¼″ x 4-¼″ tile		190	.084		1.25	1.52		2.77	3.53	
	5500	6″ x 4-¼″ tile		190	.084		1.60	1.52		3.12	3.91	
	5700	8-½″ x 4-¼″ tile		190	.084		2.20	1.52		3.72	4.57	
	5800	6″ x 6″ tile		200	.080		1.75	1.44		3.19	3.97	
	6000	Decorated wall tile, 4-¼″ x 4-¼″, minimum		870	.018	Ea.	.80	.33		1.13	1.35	
	6100	Maximum		580	.028	″	9.50	.50		10	11.15	
	6300	Exterior walls, frostproof, mud set, 4-¼″ x 4-¼″		102	.157	S.F.	4	2.83		6.83	8.40	
	6400	1-⅜″ x 1-⅜″		93	.172		7.50	3.11		10.61	12.65	
	6600	Crystalline glazed, 4-¼″ x 4-¼″, mud set, plain		100	.160		1.85	2.89		4.74	6.10	
	6700	4-¼″ x 4-¼″, scored tile		100	.160		2.05	2.89		4.94	6.35	
	6900	1-⅜″ squares		93	.172		4.25	3.11		7.36	9.05	
	7000	For epoxy grout, 1/16″ joints, 4-¼″ tile, add		800	.020		.40	.36		.76	.95	
	7200	For tile set in dry mortar, add		1,735	.009			.17		.17	.24	
	7300	For tile set in portland cement mortar, add		290	.055			1		1	1.41	
	9000	Regrout tile 4-½ x 4-½, or larger, wall	1 Tilf	100	.080		.11	1.60		1.71	2.38	
	9220	Floor	″	125	.064		.11	1.28		1.39	1.93	
100	0010	**CERAMIC TILE PANELS** Insulated, over 1000 S.F., 1-½″ thick	D-7	220	.073		6.60	1.31		7.91	9.10	100
	0100	2-½″ thick		220	.073		7.20	1.31		8.51	9.75	
150	0010	**GLASS MOSAICS** ¾″ tile on 12″ sheets, color group 1 & 2, min.		82	.195		8	3.52		11.52	13.80	150
	0300	Maximum (latex set)		73	.219		9.50	3.96		13.46	16.05	
	0350	Color group 3		73	.219		10	3.96		13.96	16.60	
	0400	Color group 4		73	.219		11.50	3.96		15.46	18.25	
	0450	Color group 5		73	.219		14.75	3.96		18.71	22	
	0500	Color group 6		73	.219		19	3.96		22.96	26	
	0600	Color group 7		73	.219		22	3.96		25.96	30	
	0700	Color group 8, golds, silvers & specialties		64	.250		33	4.51		37.51	43	

108

Figure 25.7

Wall and Ceiling Finishes

Wall and ceiling finishes are available in various types of material combinations and colors. The majority of these items will be specified in Division 9 – Finishes. Unique items, such as mirrors, may be specified in Division 8 – Doors, Windows & Glass. Items such as corner guards may be found in Division 5 – Metals.

The type of wallcovering or ceiling material and pattern repeat, if any, will significantly impact the quantity required. For example, when the wall covering is patterned and the walls have a series of inside and outside corners, up to 25% additional material may have to be added to ensure a proper match. The proper subsurface should be applied (or prepared) before the wall covering is installed or paint is applied. The estimate should include any additional work required, to prepare the walls in accordance with manufacturers' recommendations/instructions.

In this estimating example, wall treatments include paint and vinyl wallcovering accent walls.

Painting

To determine the amount of painting required, the room finish schedule, drawings, and specifications should first be reviewed. These documents will specify the surface preparation, type, and number of coats of paint, and other application details. Each room should be treated individually. Painting is measured by the square foot. If different types of paint (e.g., flat and enamel) are specified, they should be listed separately. Number of coats for each type should also be listed. The area of openings is usually not deducted, if less than 100 square feet. Painting quantities for items such as trim or grilles can be determined as equivalent wall area, as shown in Chapter 9, "Finishes", Figure 9.11. Costs for painting for the sample estimate are shown in Figure 25.8.

Costs for these items are taken from Means *Interior Cost Data*, 1987, Division 9.8 – Painting and Wall Covering (Figure 25.9). Note that the application method has a significant impact on cost.

Wallcovering

Different methods for determining quantities for wallcovering are discussed in Chapter 9, "Finishes". In the example, one accent wall will have vinyl wallcovering. The specified goods are 54" wide. The wall to be covered is 11'-6" wide by 8' high, or 92 S.F.. This quantity (100 S.F., including waste) can be used for pricing. However, to determine the quantity of material to be purchased, the estimator must consider the width of the goods. In this case, three strips (54" wide, approximately 8'-6" high) are required. A bolt 25'-6", or 8.5 yards long is required for each suite. The supplier must provide 115 S.F. of material for each suite. This is compared to the 100 S.F. area used for pricing.

Ceilings

Finish ceiling materials will be specified in Division 9 and on the Room Finish Schedule. Cost estimates should include special requirements such as returns or soffits required for boxing around mechanical and electrical equipment and unique lighting fixtures. The specifications should include provisions for appropriate surface preparation. Ceilings are most often measured by the square foot.

In the hotel suite example, the living room and bedroom ceilings are to be spray painted (texture). The kitchen, bathroom, and closet receive a suspended acoustical ceiling for concealment of piping and mechanical equipment. Costs are shown in Figure 25.10. Costs for the suspended ceiling are increased by 25% because the rooms are small.

Means Forms

COST ANALYSIS

PROJECT: Hotel Suite - Plan A ESTIMATE NO. 87-1

ARCHITECT DATE 1987

TAKE OFF BY: FWH QUANTITIES BY: FWH PRICES BY: RSm EXTENSIONS BY: SLP CHECKED BY: JDm

DESCRIPTION	SOURCE/DIMENSIONS			QUANTITY	UNIT	MATERIAL		LABOR		EQ./TOTAL	
						UNIT COST	TOTAL	UNIT COST	TOTAL	UNIT COST	TOTAL
Wall Finishes:											
Primer- all walls											
Living Room				440	SF						
Bedroom } latex				320	SF						
Closet				88	SF						
Bathroom } enamel				208	SF						
Kitchen				168	SF						
	9.8	210	0240	1224	SF	04	50	07	86		
2 coats flat latex	9.8	210	0840	756	SF	09		14	106		
(material deduct)	9.8	210	1700	(- 10%)		08	60				
2 coats enamel	9.8	210	0840	376	SF	09	34	14	53		
Doors - 5 doors	9.8	170	1200	10	Ea.	2⁵⁵	25	21	210		
2 coats per side											
Vinyl Wallcovering	9.8	400	3300	100	SF	61	61	33	33		
Accent Wall											
Total							230		488		

Figure 25.8

9.8 Painting & Wall Covering		CREW	DAILY OUTPUT	MAN-HOURS	UNIT	BARE COSTS				TOTAL INCL O&P
						MAT.	LABOR	EQUIP.	TOTAL	
210	0100	Concrete, dry wall or plaster, oil base, primer or sealer coat									210
	0200	Smooth finish, brushwork	1 Pord	1,900	.004	S.F.	.04	.08		.12	.16
	0240	Roller		2,200	.004		.04	.07		.11	.15
	0280	Spray		5,000	.002		.04	.03		.07	.09
	0300	Sand finish, brushwork		1,700	.005		.05	.09		.14	.19
	0340	Roller		2,100	.004		.05	.07		.12	.16
	0380	Spray		3,750	.002		.05	.04		.09	.12
	0400	Paint 1 coat, smooth finish, brushwork		1,800	.004		.05	.09		.14	.18
	0440	Roller		2,100	.004		.05	.07		.12	.16
	0480	Spray		3,750	.002		.05	.04		.09	.12
	0500	Sand finish, brushwork		1,600	.005		.06	.10		.16	.21
	0540	Roller		2,000	.004		.06	.08		.14	.18
	0580	Spray		3,750	.002		.06	.04		.10	.13
	0800	Paint 2 coats, smooth finish, brushwork		975	.008		.09	.16		.25	.33
	0840	Roller		1,125	.007		.09	.14		.23	.30
	0880	Spray		2,250	.004		.09	.07		.16	.20
	0900	Sand finish, brushwork		825	.010		.11	.19		.30	.39
	0940	Roller		1,050	.008		.11	.15		.26	.34
	0980	Spray		2,250	.004		.11	.07		.18	.22
	1200	Paint 3 coats, smooth finish, brushwork		675	.012		.13	.23		.36	.48
	1240	Roller		790	.010		.13	.20		.33	.43
	1280	Spray		1,500	.005		.13	.10		.23	.29
	1300	Sand finish, brushwork		560	.014		.16	.28		.44	.58
	1340	Roller		710	.011		.16	.22		.38	.49
	1380	Spray	▼	1,500	.005	▼	.16	.10		.26	.33
	1500										
	1600	Glaze coating, 5 coats, spray, clear	1 Pord	900	.009	S.F.	.50	.17		.67	.80
	1640	Multicolor	"	900	.009		.60	.17		.77	.91
	1700	For latex paint, deduct					10%				
	1800	For ceiling installations, add				▼		25%			
	1900										
	2000	Masonry or concrete block, oil base, primer or sealer coat									
	2100	Smooth finish, brushwork	1 Pord	1,725	.005	S.F.	.06	.09		.15	.20
	2180	Spray		3,750	.002		.06	.04		.10	.13
	2200	Sand finish, brushwork		1,400	.006		.07	.11		.18	.24
	2280	Spray		3,750	.002		.07	.04		.11	.14
	2400	Paint 1 coat, smooth finish, brushwork		1,550	.005		.06	.10		.16	.21
	2480	Spray		3,750	.002		.06	.04		.10	.13
	2500	Sand finish, brushwork		1,140	.007		.07	.14		.21	.27
	2580	Spray		3,750	.002		.07	.04		.11	.14
	2800	Paint 2 coats, smooth finish, brushwork		1,000	.008		.10	.16		.26	.34
	2880	Spray		2,250	.004		.10	.07		.17	.21
	2900	Sand finish, brushwork		666	.012		.11	.23		.34	.46
	2980	Spray		2,250	.004		.11	.07		.18	.22
	3200	Paint 3 coats, smooth finish, brushwork		727	.011		.15	.22		.37	.48
	3280	Spray		1,500	.005		.15	.10		.25	.32
	3300	Sand finish, brushwork		470	.017		.16	.33		.49	.66
	3380	Spray		1,500	.005		.16	.10		.26	.33
	3600	Glaze coating, 5 coats, spray, clear		900	.009		.50	.17		.67	.80
	3620	Multicolor		900	.009		.60	.17		.77	.91
	4000	Block filler, 1 coat, brushwork		1,350	.006		.08	.12		.20	.26
	4100	Silicone, water repellent, 2 coats, spray	▼	900	.009		.04	.17		.21	.29
	4120	For latex paint, deduct				▼	10%				
220	0010	REMOVAL Existing lead paint, by chemicals,									220
	0020	refinish with 2 coats of paint									
	0050	Baseboard, to 6" wide	1 Pord	190	.042	L.F.	.13	.82		.95	1.33
	0070	To 12" wide		150	.053	"	.26	1.04		1.30	1.79
	0200	Balustrades, one side	▼	90	.089	S.F.	.29	1.74		2.03	2.83
	0220										

121

Figure 25.9

Means Forms

COST ANALYSIS

SHEET NO. 3 of 4

PROJECT Hotel Suite- Plan A

ARCHITECT

ESTIMATE NO. 87-1

DATE 1987

TAKE OFF BY: FWH   QUANTITIES BY: FWH   PRICES BY: RSM   EXTENSIONS BY: SLP   CHECKED BY: JDM

DESCRIPTION	SOURCE/DIMENSIONS			QUANTITY	UNIT	MATERIAL		LABOR		EQ./TOTAL	
						UNIT COST	TOTAL	UNIT COST	TOTAL	UNIT COST	TOTAL
Ceilings:											
Living Room				184	SF						
Bedroom				96	SF						
Prime	9.8	210	0240	280	SF	04	11	07	20		
Sand coat	9.8	210	0540	280	SF	06	17	08	22		
Suspended Acoustical											
Bathroom				40	SF						
Closet				15	SF						
Kitchen				28	SF						
	9.5	250	0800	83	SF	1.13	94	48	40		
Add for small area	9.5	250	2500				25%		10		
Total							(122)		(92)		

Figure 25.10

242

The Estimate Summary

Once the costs for material and labor have been determined for each type of work, an estimate summary is completed (shown in Figure 25.11). Percentages for indirect costs can be added at this stage or when costs for all hotel suites have been totalled. For this example, markups will be included for one suite. The following costs could be added at the estimate summary stage:

- General Conditions
- Sales Tax
- Overhead and Profit
- Contingency
- Design Fees

Discussions of these costs and their application are included in Chapters 21, 22, and 24.

Furnishings

Furnishings include all movable items in the sample hotel suite. The furnishings specification package is generally separate from the construction documents, and is usually highly detailed. Furniture plans generally include a coding system, as shown in Figure 25.12. It is from this coding that the specific manufacturer, product number, style, color, and other details may be referenced. A furnishings design package will include drawings, specifications, and possibly color boards with catalogue cuts and material samples. Special installation notes will be included where appropriate. Architectural items which might be part of a furnishings package (e.g., bulletin boards, blinds, or architectural woodwork) may be listed elsewhere, for example in Division 10 – Specialties or Division 6 – Wood and Plastics. The contract documents should be thoroughly read for any architectural items or other requirements which may not be listed in the furnishings documents.

Furnishings should be counted and listed systematically, starting with one furnishing type and style, and proceeding to the next, room by room. A typical listing is as shown below for the hotel suite estimating example. The coding from the plan is used to reference a more detailed code which can identify the manufacturer, product number, style, etc.

	CODE	DESCRIPTION	QTY
Living Room:			
C1	Chair 1 – 11-249	Lounge Chair	2
C2	Chair 2 – 8-565	Sleep Sofa (Full Size)	1
C3	Chair 3 – 4614	Desk Chair	1
T1	Table 1 – 2717-651	Cocktail Table	1
T2	Table 2 – 2717-601	End Table	2
T3	Table 3 –	TV Stand and TV	1
D1	Desk 1 – 546-18		1
A1	Armoire 1 – 546-465		1
L1	Lamp 1	Desk lamp	1
L2	Lamp 2	Floor lamp	1
L3	Lamp 3	Table lamps	2

CODE	DESCRIPTION	QTY
Living Room: (Continued)		
P1	Plant 1	
WG1	Wall Graphic 1	1
WG2	Wall Graphic 2	1
WG3	Wall Graphic 3	1
WG4	Wall Graphics 4	1
WB1	1" Slat Window Blinds 3' x 5'	15 S.F.
Bedroom:		
DU1	Drawer Unit — 2317-281	1
M1	Mirror — 546-231	1
L4	Lamp 4	Floor lamp	1
L5	Lamp 5	Wall-mount lamp	1
BF1	Bed frame	1
BSM1	Box spring and Mattress	1
HDB1	Headboard — 2317-3651	1
WG5	Wall Graphic 5	1
WG6	Wall Graphic 6	1
WG7	Wall Graphic 7	1
BS1	Bedspread	2
WB1	1" Slat Window Blinds 3' x 5'	15 S.F.

Typical furnishings costs are listed in Means *Interior Cost Data*, 1987, Division 12.1 — Furnishings, and for this example, under Furniture, Hotel, (shown in Figure 25.13). Minimum and maximum costs are often listed. If, in the conceptual stage of the project, it is known that the project will be "high end" in design, the maximum cost should be used. If it will be a "low end" project, the minimum cost should be used, and for an average project, an average cost should be used.

If furniture is specified by manufacturer, product number, and finish, the estimate can be very accurate. Costs for each item can be obtained from the specific manufacturer or supplier. If manufacturer, product number, and specific finishes are not identified, costs listed in *Interior Cost Data* can be used to develop the estimate. All furnishings costs in the book are based on commercial furnishings used in today's hotels, restaurants, office buildings, and other public and institutional facilities. It is assumed the example is a high end project, therefore, the maximum costs will be used. Costs for furniture are entered on the Cost Analysis sheets shown in Figure 25.14 and Figure 25.15 and are net costs.

Once all furnishing costs are determined, the costs can be transferred to a summary sheet, shown in Figure 25.16. Costs for dealer or supplier profit, sales tax, packing, delivery, storage, and installation should be included if required.

COST ANALYSIS

PROJECT Hotel Suite - Plan A

ESTIMATE NO. 87-1

ARCHITECT

DATE 1987

TAKE OFF BY: FWH   QUANTITIES BY: FWH   PRICES BY: RSM   EXTENSIONS BY: SLP   CHECKED BY: JDM

DESCRIPTION	SOURCE/DIMENSIONS			QUANTITY	UNIT	MATERIAL		LABOR		EQ./TOTAL	
						UNIT COST	TOTAL	UNIT COST	TOTAL	UNIT COST	TOTAL
Estimate Summary											
Floor Finishes							870		251		1121
Wall Finishes							230		488		718
Ceilings							122		92		214
Subtotal							1222		831		2053
General Conditions	15%						183		125		308
Sales Tax - 5% on mat.							70				70
Subtotal							1475		956		2431
Overhead & Profit - 10% mat.	47.6% Labor						147		455		602
Subtotal							1622		1411		3033
Contingency - 10%							162		141		303
Total							1784		1552		(3336)

Figure 25.11

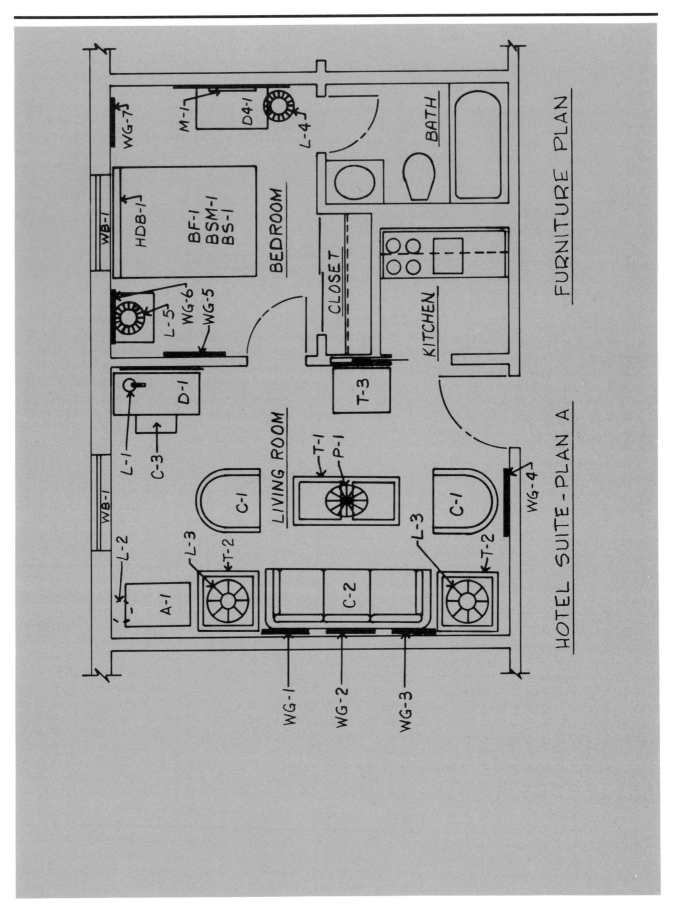

WG-7
M-I
D4-I
L-4
BATH

WB-I
HDB-I
BF-I
BSM-I
BS-I
BEDROOM

L-5
WG-6
WG-5
CLOSET
KITCHEN

D-I
L-I
C-3

WB-I
L-2
L-3
T-2
C-I
LIVING ROOM
T-I
P-I
C-I
L-3
T-2

A-I
C-2
WG-4

WG-I
WG-2
WG-3

T-3

FURNITURE PLAN

HOTEL SUITE - PLAN A

Figure 25.12

12.1 Furnishings	CREW	DAILY OUTPUT	MAN-HOURS	UNIT	BARE COSTS MAT.	LABOR	EQUIP.	TOTAL	TOTAL INCL O&P	
450 0300 Manual and electric beds, minimum				Ea.	650			650	715	**450**
0400 Maximum					1,400			1,400	1,550	
0600 All electric hospital beds, minimum					865			865	950	
0700 Maximum					2,150			2,150	2,375	
0900 Manual, nursing home beds, minimum					410			410	450	
1000 Maximum					850			850	935	
1020 Overbed table, laminated top, minimum					175			175	195	
1040 Maximum					435			435	480	
1100 Patient wall systems, not incl. plumbing, minimum				Room	520			520	570	
1200 Maximum				"	960			960	1,050	
2000 Geriatric chairs, minimum				Ea.	190			190	210	
2020 Maximum				"	355			355	390	
3002 Hospital furniture-see also division 12.1-500										
3010										
3022 Hospital mattress-see also division 12.1-660										
3030										
3042 Hospital seating-see also division 12.1-500										
500 0010 **FURNITURE, HOTEL** Standard quality set, minimum				Room	1,675			1,675	1,850	**500**
0200 Maximum				"	5,150			5,150	5,675	
0300 Bed frame				Ea.	29			29	32	
0350										
0400 Bench, upholstered, 42" x 18" x 18"				Ea.	112			112	125	
0420 18" x 18" x 23"					61			61	67	
0500 Desk section, one drawer, 34" x 20" x 30"					90			90	99	
0600 Free standing, 42" x 22" x 30"					138			138	150	
0700 Desk chair, upholstered, minimum					73			73	80	
0720 Maximum					175			175	195	
1000 Dressers, uniplex, 2 drawer					299			299	330	
1100 3 drawer					350			350	385	
2000 Guest tables, 30" diameter					127			127	140	
2100 34" diameter					151			151	165	
3000 Headboards, free standing, twin					57			57	63	
3050 Full					108			108	120	
3100 Queen					108			108	120	
3200 Wall mounted, twin					60			60	66	
3250 Full					102			102	110	
3300 Queen					109			109	120	
4000 Lounge chair, full upholstered, minimum					155			155	170	
4050 Maximum					531			531	585	
4200 Open arms, minimum					85			85	94	
4250 Maximum					174			174	190	
5000 Mattress/box springs, twin					144			144	160	
5050 Full					175			175	195	
5100 Queen					227			227	250	
5150 Mirror, framed, 29" x 45"					75			75	83	
6000 Sleep sofas, twin, minimum					335			335	370	
6050 Maximum					531			531	585	
6100 Full, minimum					366			366	405	
6150 Maximum					716			716	790	
6200 Queen, minimum					402			402	440	
6250 Maximum					726			726	800	
7000 Table, wood top, cocktail, 54" x 24" x 16" high					124			124	135	
7020 Corner, 30" x 30" x 21" high					113			113	125	
7040 End, 22" x 28" x 21" high					103			103	115	
510 0010 **FURNITURE, LIBRARY**										**510**
0100 Attendant desk, 36" x 62" x 29" high	1 Carp	16	.500	Ea.	1,430	10.30		1,440	1,600	
0200 Book display, "A" frame display, both sides		16	.500		1,612	10.30		1,622	1,800	
0220 Table with bulletin board		16	.500		853	10.30		863.30	955	

154

Figure 25.13

Means Forms

COST ANALYSIS

PROJECT Hotel Suite - Plan A

ARCHITECT

TAKE OFF BY: AEL   QUANTITIES BY: AEL   PRICES BY: RSM   EXTENSIONS BY: SLP   CHECKED BY: JDM

Furnishings

SHEET NO. 1 of 3

ESTIMATE NO. 87-1a

DATE 1987

DESCRIPTION		SOURCE/DIMENSIONS			QUANTITY	UNIT	MATERIAL		LABOR		EQ./TOTAL	
							UNIT COST	TOTAL	UNIT COST	TOTAL	UNIT COST	TOTAL
Living Room												
Chair	C1	12.1	500	4050	2	Ea.	531	1062				
Sleep Sofa	C2	12.1	500	6150	1	Ea.	716	716				
Desk Chair	C3	12.1	500	0720	1	Ea.	175	175				
Cocktail Table	T1	12.1	500	7000	1	Ea.	124	124				
End Table	T2	12.1	500	7040	2	Ea.	103	206				
TV Table	T3	12.1	500	7040	1	Ea.	103	103				
Desk	D1	12.1	500	0600	1	Ea.	138	138				
Armoire	A1	12.1	420	7000	1	Ea.	324	324				
Desk Lamp	L1	12.1	640	3100	1	Ea.	105	105				
Floor Lamp	L2	12.1	640	3450	1	Ea.	180	180				
Table Lamp	L3	12.1	640	1420	2	Ea.	77	154				
Wall Graphics	WG1	12.1	030	3050	1	Ea.	96⁵⁰	97				
	WG2	12.1	030	3050	1	Ea.	96⁵⁰	97				
	WG3	12.1	030	3050	1	Ea.	96⁵⁰	97				
	WG4	12.1	030	3050	1	Ea.	96⁵⁰	97				
Blinds	WB1	12.1	100	0100	15	SF	5⁶⁷	84				
								(3759)				

Figure 25.14

248

Means Forms

COST ANALYSIS

PROJECT: Hotel Suite - Plan A

ARCHITECT:

TAKE OFF BY: AEL   QUANTITIES BY: AEL   PRICES BY: RSM   EXTENSIONS BY: SLP   CHECKED BY: JDM

Furnishings

SHEET NO. 2 of 3

ESTIMATE NO. 87-1a

DATE 1987

DESCRIPTION		SOURCE/DIMENSIONS			QUANTITY	UNIT	MATERIAL UNIT COST	MATERIAL TOTAL	LABOR UNIT COST	LABOR TOTAL	EQ./TOTAL UNIT COST	EQ./TOTAL TOTAL
Bedroom												
Dresser	DU 1	12.1	500	1100	1	Ea.	350	350				
Mirror	M1	12.1	500	5150	1	Ea.	75	75				
Floor Lamp	L4	12.1	640	3450	1	Ea.	180	180				
Wall Lamp	L5	12.1	640	3350	1	Ea.	113	113				
Bed Frame	BF1	12.1	500	0300	1	Ea.	29	29				
Box Spring & mattress	BSM1	12.1	500	5050	1	Ea.	175	175				
Headboard	HDB1	12.1	500	3050	1	Ea.	108	108				
Wall Graphics	WG 5	12.1	030	3050	1	Ea.	96⁵⁰	97				
	WG 6	12.1	030	3050	1	Ea.	96⁵⁰	97				
	WG 7	12.1	030	3050	1	Ea.	96⁵⁰	97				
Bed Spread	BS1	12.1	250	8500	2	Ea.	221	442				
Blinds	WB1	12.1	100	0100	15	SF	5⁵⁷	84				
								(1847)				

Figure 25.15

249

Means Forms

COST ANALYSIS

PROJECT **Hotel Suite — Plan A**

ESTIMATE NO. **87- 1a**

ARCHITECT

DATE **1987**

TAKE OFF BY: **AEL**   QUANTITIES BY: **AEL**   PRICES BY: **RSM**   EXTENSIONS BY: **SLP**   CHECKED BY: **JDM**

DESCRIPTION	SOURCE/DIMENSIONS			QUANTITY	UNIT	MATERIAL		LABOR		EQ./TOTAL	
						UNIT COST	TOTAL	UNIT COST	TOTAL	UNIT COST	TOTAL
Furnishings Estimate Summary											
Living Room							3759				
Bedroom							1847				
Subtotal	(net)						5606				
Dealer markup	50%						2803				
Subtotal							8409				
Sales Tax	5%						420				
Packing	7%						588				
Delivery	10%						841				
Installation	5%						420				
Total							(10678)				

Figure 25.16

To determine some of these costs, the following percentages may be used:

Dealer/supplier Profit:	20% to 100% varies considerably based on quantity and other factors.
Packing and Crating:	6%–15% of the furniture cost.
Delivery Costs:	7%–25% of the furniture cost.
Installation:	0% to 25% of the furniture value, depending on project size and degree of complexity. A direct quote from an installation company or furniture dealer is the most accurate installation estimate.

The estimate summaries shown in Figure 25.11 for finishes, and Figure 25.16 for furnishings, may be added together for the total project cost of the hotel suite. As previously stated, this cost should be multiplied by the number of identical suites to be remodeled in an entire hotel project.

Using a systematic method to develop interior costs, each interior finish and furnishing item can be estimated accurately. Since the coding is easy to translate, the estimator can quickly check to ensure that all items are included in the estimate and an accurate determination of the total interior costs can be made. Often, when preliminary budgets are being developed, not all the details are known, therefore, a unit cost estimate cannot be developed. In these cases, the systems estimating method would be more appropriate. The next chapter describes the systems method, and illustrates this method through an estimating example.

Chapter 26

SYSTEMS ESTIMATING EXAMPLE

The Systems Estimate is useful during the design development stage of a project. The estimator needs only basic project parameters and perhaps a preliminary floor plan to complete the estimate effectively. The advantage of using a Systems Estimate is the ability to develop costs quickly and to establish a budget before preparation of working drawings and specifications. The estimator can easily substitute one system for another to determine the most cost effective approach. The Systems Estimate can be completed in much less time than the Unit Price Estimate. Some accuracy is sacrificed, however, and the Systems Estimate should be used only for budgetary purposes.

Budgets and cost control are becoming increasingly more important before the project enters the final design process – when owners take on the expense of working drawings and specifications. It is crucial that the estimator combine a thorough evaluation of the existing conditions with the design parameters in order to properly complete the Systems Estimate. The estimator must use experience to be sure to include all requirements since little information is provided. Applicable building codes and fire codes and local regulations must also be considered; requirements can significantly affect cost and restrictions may dictate design parameters.

Prices and costs used in the following example are from Means *Interior Cost Data*, 1987. A description of the use of the Systems pages is included in Chapter 24 of this book. The sample project below may vary in detail from actual projects. Every interior project should be treated individually.

Project Description

The example project is the renovation of a space within a turn-of-the-century mill building into commercial office space. A preliminary floor plan is shown in Figure 26.1. The area to be estimated is highlighted. Exterior walls are sandblasted brick with new windows. The partitions, entrance doors, elevators, and bathrooms are in place. Heat (hot water) is provided to the space for distribution by the tenant. This is typical of the type of information available to the estimator preparing a preliminary budget.

Though only at the budgetary stage, it is important that the interior estimator visit the site of the proposed work, whether the space is in a new building or to be renovated. Existing conditions cannot be adequately described in written or verbal communication. Such conditions can have an impact on the cost.

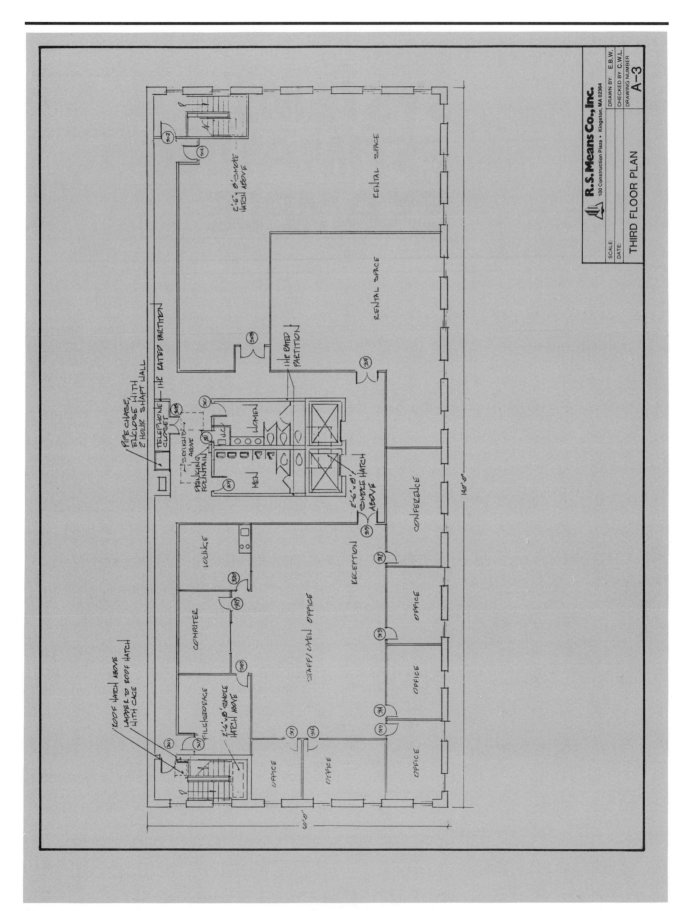

Figure 26.1

Preparing the Estimate

When performing an estimate for budgetary purposes, the use of systems or assemblies costs is most appropriate because the estimate can be relatively accurate and can be completed in a short amount of time. Realistically, every possible type of work may not be defined as a system. In such cases, unit prices are used. In this example, both systems costs and unit prices are included. Since the systems costs include overhead and profit, the unit costs, when used, must also include these mark-ups. Each type of cost will be designated (S) or (U) when entered on the estimate sheets.

Partitions and Doors

Components of the drywall partitions are taken from the table in Figure 26.2. This table allows for almost endless variation of drywall partition systems. In this case, the estimator has chosen 5/8" drywall on metal studs with fiberglass insulation (for sound deadening). Costs are entered on the estimate sheet shown in Figure 26.3. Note that a cost per square foot of partition and a cost per square foot of floor area can be developed for the complete system. Both of these costs can be used for future estimating.

For certain items, costs may not be available for the exact type and size of the component required. Interpolation from the costs of similar items can provide an accurate budget amount. In the example, costs for the doors are determined in this manner, as shown in Figures 26.3 and 26.4. These costs include finishing for the doors and frames.

Finishes

Wall, floor, and ceiling finishes for the sample project are shown in Figure 26.5. Most of the costs are taken from the unit price pages of Means *Interior Cost Data*, 1987.

For painting, two items (primer and finish) are added to obtain a total square foot price. This cost can then be compared to the cost for vinyl wallcovering. By making such cost comparisons, the designer, owner, or estimator is able to make design decisions based on budget restrictions *prior* to preparation of working drawings and the solicitation of bids. In other words, the correct choice is made before time and money are spent. Similar decisions can be made for the other finishes.

Mechanical and Electrical

In order to properly determine budget costs for the mechanical and electrical portion of a building, prior to engineering and drawings, the estimator should be somewhat familiar with the work and local codes and requirements. When the type of system has been chosen, the best source of costs for the system is from past, similar projects. However, the estimator must be careful because types of HVAC systems (heating, ventilation, and air conditioning) used in similar types of spaces (e.g., offices) can vary significantly in design and in cost. The type of heating and/or cooling generation system, if in place in the building, will also dictate which type of system is most appropriate.

For the example project heat, in the form of hot water, is provided to the space. Options for heat distribution include baseboard or a hot water coil in ductwork. Since the space will be air conditioned (through ductwork) the latter option is the most cost effective. Costs for the system are interpolated from data in Figure 26.6 and entered on the estimate sheet in Figure 26.7.

PARTITIONS	B6.1-580	Drywall Components

6.1-580	Drywall Components	COST PER S.F.		
		MAT.	INST.	TOTAL
0140	Metal studs, 24″ O.C. including track, load bearing, 18 gage, 2-½″	.72	.48	1.20
0160	3-⅝″	.75	.50	1.25
0180	4″	.79	.54	1.33
0200	6″	.97	.55	1.52
0220	16 gage, 2-½″	.87	.50	1.37
0240	3-⅝″	.92	.54	1.46
0260	4″	.95	.54	1.49
0280	6″	1.13	.58	1.71
0300	Non load bearing, 25 gage, 1-⅝″	.17	.48	.65
0340	3-⅝″	.21	.50	.71
0360	4″	.25	.54	.79
0380	6″	.31	.55	.86
0400	20 gage, 2-½″	.32	.48	.80
0420	3-⅝″	.39	.50	.89
0440	4″	.44	.54	.98
0460	6″	.54	.55	1.09
0540	Wood studs including blocking, shoe and double top plate, 2″x4″, 12″O.C.	.38	.63	1.01
0560	16″ O.C.	.32	.51	.83
0580	24″ O.C.	.25	.41	.66
0600	2″x6″, 12″ O.C.	.58	.72	1.30
0620	16″ O.C.	.49	.56	1.05
0640	24″ O.C.	.39	.44	.83
0642	Furring one side only, steel channels, ¾″, 12″ O.C.	.31	.99	1.30
0644	16″ O.C.	.24	.88	1.12
0646	24″ O.C.	.20	.67	.87
0647	1-½″ , 12″ O.C.	.47	1.12	1.59
0648	16″ O.C.	.37	.98	1.35
0649	24″ O.C.	.31	.76	1.07
0650	Wood strips, 1″ x 3″, on wood, 12″ O.C.	.10	.44	.54
0651	16″ O.C.	.08	.33	.41
0652	On masonry, 12″ O.C.	.11	.49	.60
0653	16″ O.C.	.08	.37	.45
0654	On concrete, 12″ O.C.	.12	.93	1.05
0655	16″ O.C.	.09	.70	.79
0660	Gypsum board, one face only, exterior sheathing, ½″	.24	.48	.72
0680	Fire resistant, ½″	.31	.27	.58
0700	⅝″	.32	.28	.60
0720	Sound deadening board ¼″	.18	.26	.44
0740	Standard drywall ⅜″	.26	.25	.51
0760	½″	.28	.26	.54
0780	⅝″	.29	.28	.57
0800	Tongue & grove coreboard 1″	.40	1	1.40
0820	Water resistant, ½″	.34	.27	.61
0840	⅝″	.37	.29	.66
0860	Add for the following:, foil backing	.12		.12
0880	Fiberglass insulation, 3-½″	.24	.15	.39
0900	6″	.37	.18	.55
0920	Rigid insulation 1″	.55	.24	.79
0940	Resilient furring @ 16″ O.C.	.14	.74	.88
0960	Taping and finishing	.02	.24	.26
0980	Texture spray	.11	.33	.44
1000	Thin coat plaster	.09	.37	.46
1040	2″x4″ staggered studs 2″x6″ plates & blocking	.37	.73	1.10
1060	1″x6″ gypsum studs	.18	.51	.69

211

Figure 26.2

256

Means Forms
COST ANALYSIS

SHEET NO. 1 of 4

PROJECT **Office Renovation - 3020 S.F.** ESTIMATE NO. 87-2

ARCHITECT **CWL** DATE 1987

TAKE OFF BY: **EBW** QUANTITIES BY: **EBW** PRICES BY: **RSM** EXTENSIONS BY: **SLP** CHECKED BY: **JDM**

	Source			Qty.	Unit		$/SF Floor		Unit Cost	Total Incl. O&P
Partitions:										
3⅝" 25 ga. studs	6.1	580	0340	3015	SF	(s)			71	2141
3½" fiberglass unfaced	6.1	580	0880	3015	SF	(s)			39	1176
⅝" std. drywall	6.1	580	0780	6030	SF	(s)			57	3437
Tape & finish	6.1	580	0960	6030	SF	(s)			26	1568
Partitions Total										(8322)
Cost /SF								(2.76)		

$$\frac{Total}{SF\ Wall} = \frac{\$8322}{3015\ SF} = \$\qquad /SF\ Wall$$

	Source			Qty.	Unit		$/SF Floor		Unit Cost	Total Incl. O&P
Doors:										
3°x 7° SC oak flush	6.4	260	5200 5360	9	Ea.	(s)			350	3150
Locksets	8.7	350	0400	9	Ea.	(u)			79	711
Hinges (material only)	8.7	330	0040	13.5	Pr.	(u)			26	351
Doors Total										(4212)
Cost / SF								(1.39)		
Glass: (Computer Room)										
Alum. tube fr., ¼" temp. 6'x 4'	6.1	700	1400	1	Ea.	(s)			515	(515)
Cost / SF								(17)		

Figure 26.3

| 6.4-240 | | | | | Wood Door/Wood Frame | | | |

	TYPE	FACE	SIZE	FRAME	DEPTH	COST EACH		
						MAT.	INST.	TOTAL
2360			6'-0" x 6'-8"	pine	3-5/8"	225	175	400
2380					5-3/16"	245	185	430
2400		plastic laminate	2'-6" x 6'-8"	pine	3-5/8"	190	115	305
2420					5 3/16"	205	125	330
2560			6'-0" x 6'-8"	pine	3-5/8"	380	170	550
2580					5-3/16"	400	180	580
3000	Particle core/flush	lauan	2'-6" x 6'-8"	pine	3-5/8"	120	115	235
3020					5-3/16"	135	120	255
3160			6'-0" x 7'-0"	pine	3-5/8"	255	180	435
3180					5-3/16"	275	190	465
3200		birch	2'-6" x 6'-8"	pine	3-5/8"	130	115	245
3220					5-3/16"	145	120	265
3360			6'-0" x 7'-0"	pine	3-5/8"	255	190	445
3380					5-3/16"	275	195	470
3400		oak	2'-6" x 6'-8"	oak	3-5/8"	145	115	260
3420					5-3/16"	165	120	285
3560			6'-0" x 7'-0"	oak	3-5/8"	285	190	475
3580					5-3/16"	305	200	505
3600		M.D. overlay	2'-6" x 6'-8"	pine	3-5/8"	125	125	250
3620		on hardboard			5-3/16"	140	135	275
3760			6'-0" x 7'-0"	pine	3-5/8"	250	200	450
3780					5-3/16"	270	210	480
3800		plastic laminate	2'-6" x 6'-8"	pine	3-5/8"	190	120	310
3820					5-3/16"	200	125	325
3960			6'-0" x 7'-0"	pine	3-5/8"	370	190	560
3980					5-3/16"	390	205	595
4400	Solid core/panel	solid pine	2'-6" x 6'-8"	pine	3-5/8"	155	120	275
4420					5-3/16"	170	130	300
4560			6'-0" x 6'-8"	pine	3-5/8"	305	195	500
4580					5-1/8"	320	205	525
5000	Solid core/flush	birch	2'-6" x 6'-8"	pine	3-5/8"	160	115	275
5020					5-3/16"	180	120	300
5160			6'-0" x 7'-0"	pine	3-5/8"	325	190	515
5180					5-3/16"	345	195	540
5200		oak	2'-6" x 6'-8"	oak	3-5/8"	180	115	295
5220					5-3/16"	195	120	315
5360			6'-0" x 7'-0"	oak	3-5/8"	350	190	540
5380					5-3/16"	370	200	570
5400		M.D. overlay	2'-6" x 6'-8"	pine	3-5/8"	160	125	285
5420		on hardboard			5-3/16"	175	135	310
5560			6'-0" x 7'-0"	pine	3-5/8"	315	200	515
5600		plastic laminate	2'-6" x 6'-8"	pine	3-5/8"	220	120	340
5620					5-3/16"	240	130	370
5760			6'-0" x 7'-0"	pine	3-5/8"	440	200	640

305

Figure 26.4

Means Forms

COST ANALYSIS

PROJECT **Office Renovation - 3020 S.F.**

ESTIMATE NO. **87-2**

ARCHITECT **CWL**

DATE **1987**

TAKE OFF BY: **EBW** QUANTITIES BY: **EBW** PRICES BY: **RSM** EXTENSIONS BY: **SLP** CHECKED BY: **JDM**

	Source			Qty.	Unit	$/SF Floor	Unit Cost	Incl. O&P
Wall Finishes:								
Painting - Primer + 2 coats	9.8	210	0240 } 0840 }	2610	SF (u)		45	1175
Sealer	9.8	210	0240	3420	SF (u)		15	513
Vinyl Wallcovering	9.8	400	3300	3420	SF (u)		1 15	3933
Wall Finishes Total								(5621)
Cost / SF						(1 86)		
Floor Finishes:								
Carpet	9.6	050	3340	222	SY (u)		25	5550
Refinish Wood (Conference)	9.6	400	7600	418	SF (u)		2 52	1053
Vinyl Composition Tile	9.6	200	7350	607	SF (u)		1 76	1068
Vinyl core base 4"	9.6	200	1150	482	LF (u)		1 21	583
Floor Finishes Total								(8254)
Cost / SF						(2 73)		
Ceiling:								
2 x 2, 3/4" reveal edge T-bar susp.	6.7	100	6100	3020	SF (s)		1 95	(5889)
Cost / SF						(1 95)		

Figure 26.5

Air Cooled Condensing Unit
Refrigerant Piping
Roof
Supply Duct
Fin. Ceiling
Return Grille — DX Air Handling Unit — Supply Diffuser

General: Split systems offer several important advantages which should be evaluated when a selection is to be made. They provide a greater degree of flexibility in component selection which permits an accurate match-up of the proper equipment size and type with the particular needs of the building. This allows for maximum use of modern energy saving concepts in heating and cooling. Outdoor installation of the air cooled condensing unit allows space savings in the building and also isolates the equipment operating sounds from building occupants.

Design Assumptions: The systems below are comprised of a direct expansion air handling unit and air cooled condensing unit with interconnecting copper tubing. Ducts and diffusers are also included for distribution of air. Systems are priced for cooling only. Heat can be added as desired either by putting hot water/steam coils into the air unit or into the duct supplying the particular area of need. Gas fired duct furnaces are also available. Refrigerant liquid line is insulated.

System Components	QUANTITY	UNIT	COST EACH		
			MAT.	INST.	TOTAL
SYSTEM 08.4-250-1280					
SPLIT SYSTEM, AIR COOLED CONDENSING UNIT					
APARTMENT CORRIDORS, 1,000 S.F., 1.80 TON					
Fan coil AC unit, cabinet mntd & filters direct expansion air cool	1.000	Ea.	245.59	60.02	305.61
Ductwork package, for split system, remote condensing unit	1.000	System	248.01	342.16	590.17
Refrigeration piping	1.000	System	70.53	368.67	439.20
Condensing unit, air cooled, incls compressor & standard controls	1.000	Ea.	650.87	239.73	890.60
TOTAL			1,215	1,010.58	2,225.58
COST PER S.F.			1.22	1.01	2.23
*Cooling requirements would lead to more than one system					

8.4-250	Split Systems With Air Cooled Condensing Units	COST PER S.F.		
		MAT.	INST.	TOTAL
1260	Split system, air cooled condensing unit			
1280	Apartment corridors, 1,000 S.F., 1.83 ton	1.22	1.01	2.23
1440	20,000 S.F., 36.66 ton	1.46	1.43	2.89
1520	Banks and libraries, 1,000 S.F., 4.17 ton	2.51	2.33	4.84
1680	20,000 S.F., 83.32 ton	3.84	3.39	7.23
1760	Bars and taverns, 1,000 S.F., 11.08 ton	6.90	4.60	11.50
1880	10,000 S.F., 110.84 ton	8.40	5.35	13.75
2000	Bowling alleys, 1,000 S.F., 5.66 ton	4.18	4.38	8.56
2160	20,000 S.F., 113.32 ton	5.45	4.79	10.24
2320	Department stores, 1,000 S.F., 2.92 ton	1.76	1.60	3.36
2480	20,000 S.F., 58.33 ton	2.32	2.28	4.60
2560	Drug stores, 1,000 S.F., 6.66 ton	4.92	5.15	10.07
2720	20,000 S.F., 133.32 ton*			
2800	Factories, 1,000 S.F., 3.33 ton	2.01	1.83	3.84
2960	20,000 S.F., 66.66 ton	3.05	2.73	5.78
3040	Food supermarkets, 1,000 S.F., 2.83 ton	1.71	1.56	3.27
3200	20,000 S.F., 56.66 ton	2.25	2.21	4.46
3280	Medical centers, 1,000 S.F., 2.33 ton	1.43	1.25	2.68
3440	20,000 S.F., 46.66 ton	1.86	1.82	3.68
3520	Offices, 1,000 S.F., 3.17 ton	1.92	1.74	3.66
3680	20,000 S.F., 63.32 ton	2.90	2.60	5.50
3760	Restaurants, 1,000 S.F., 5.00 ton	3.69	3.87	7.56
3920	20,000 S.F., 100.00 ton	4.80	4.22	9.02
4000	Schools and colleges, 1,000 S.F., 3.83 ton	2.31	2.14	4.45
4160	20,000 S.F., 76.66 ton	3.51	3.13	6.64

343

Figure 26.6

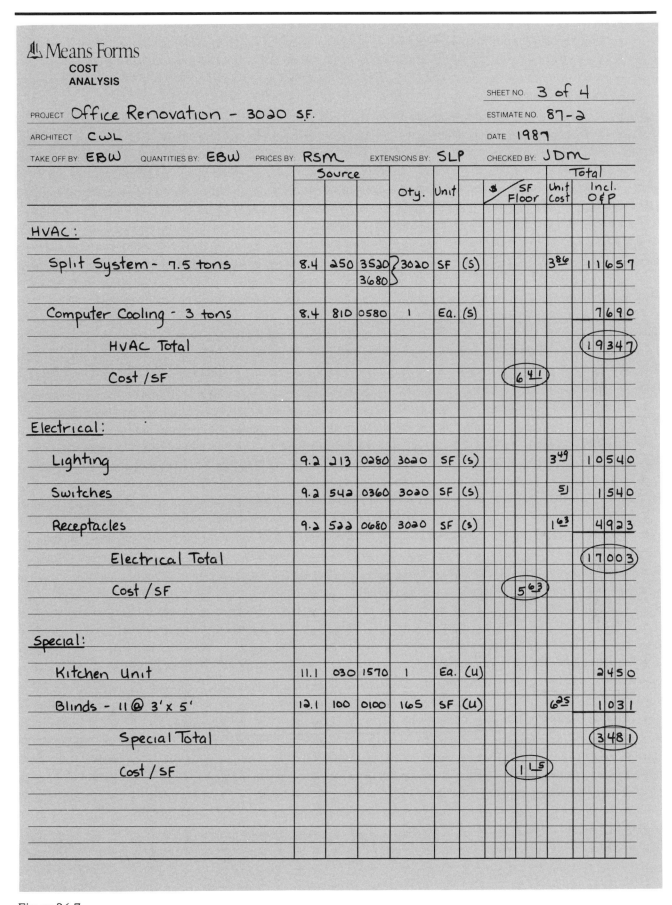

Means Forms

COST ANALYSIS

PROJECT Office Renovation - 3020 SF.

ARCHITECT CWL

DATE 1987

TAKE OFF BY: EBW   QUANTITIES BY: EBW   PRICES BY: RSM   EXTENSIONS BY: SLP   CHECKED BY: JDM

	Source			Qty.	Unit		$ / SF Floor	Unit Cost	Total Incl. O&P
HVAC:									
Split System - 7.5 tons	8.4	250	3520 } 3680	3020	SF	(s)		3⁸⁶	1 1657
Computer Cooling - 3 tons	8.4	810	0580	1	Ea.	(s)			7690
HVAC Total									(19347)
Cost /SF							(6⁴¹)		
Electrical:									
Lighting	9.2	213	0280	3020	SF	(s)		3⁴⁹	10540
Switches	9.2	542	0360	3020	SF	(s)		5¹	1540
Receptacles	9.2	522	0680	3020	SF	(s)		1⁶³	4923
Electrical Total									(17003)
Cost /SF							(5⁶³)		
Special:									
Kitchen Unit	11.1	030	1570	1	Ea.	(u)			2450
Blinds - 11 @ 3' x 5'	12.1	100	0100	165	SF	(u)		6²⁵	1031
Special Total									(3481)
Cost /SF							(1⁴⁵)		

Figure 26.7

Note that on the floor plan (Figure 26.1), in addition to office space, a computer room is included. The estimator must determine cooling requirements for such equipment because often a dedicated system is necessary. In this case, costs for a separate system are included.

The electrical portion of the estimate is a count of each type of device and fixture. Without drawings, the estimator should consult local authorities to assure compliance with applicable electrical and energy codes. With increasing energy conservation, the estimator must consider electrical usage in the estimate. Costs may be determined based on wattage per square foot as shown in Figure 26.8. Budget costs for the electrical portion of the sample project are shown on the estimate sheet in Figure 26.7.

Summary

The summary sheet for the systems estimate of the office renovation is shown in Figure 26.9. For all major components of the project, costs per square foot of floor area are developed. These are included on the summary in addition to the total costs.

The estimator can use these square foot costs as a check, to assure that the costs are "within the ball park". For example, from past similar projects, partitions normally cost about $3.50 per square foot of floor area. In the sample project, the cost is $2.76. This discrepancy should alert the estimator to check the estimate to determine why the difference exists. In this case, the exterior walls of the office space are existing sandblasted brick, therefore, no costs are included for exterior walls. Cross checking against historical data can help, not only to be sure all items are included, but also to help prevent simple mathematical errors.

Finally, because not all details concerning the project are known, a contingency is added. Suggested percentages, based on the stage of design development, are shown in Figure 26.10.

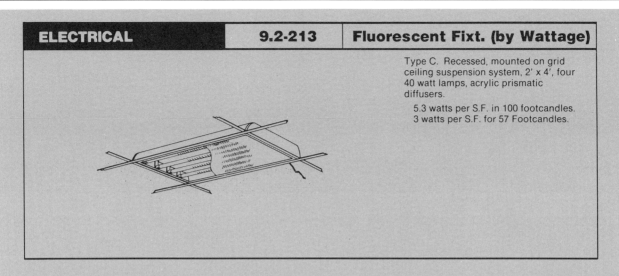

Type C. Recessed, mounted on grid ceiling suspension system, 2' x 4', four 40 watt lamps, acrylic prismatic diffusers.

5.3 watts per S.F. in 100 footcandles.
3 watts per S.F. for 57 Footcandles.

System Components			COST PER S.F.		
	QUANTITY	UNIT	MAT.	INST.	TOTAL
SYSTEM 09.2-213-0200					
FLUORESCENT FIXTURES RECESS MOUNTED IN CEILING					
1 WATT PER S.F., 20 FC, 5 FIXTURES PER 1000 S.F.					
Steel intermediate conduit, (IMC) ½" diam	.128	L.F.	.08	.34	.42
Wire, 600 volt, type THW, copper, solid, #12	.003	C.L.F.	.01	.07	.08
Fluorescent fixture, recessed, 2'x4', four 40W, w/lens, for grid ceiling	.005	Ea.	.29	.29	.58
Steel outlet box 4" square	.005	Ea.	.01	.07	.08
Fixture whip, Greenfield w/#12 THHN wire	.005	Ea.		.02	.02
TOTAL			.39	.79	1.18

9.2-213	Fluorescent Fixtures (by Wattage)	COST PER S.F.		
		MAT.	INST.	TOTAL
0190	Fluorescent fixtures recess mounted in ceiling			
0200	1 watt per S.F., 20 FC, 5 fixtures per 1000 S.F.	.39	.79	1.18
0240	2 watts per S.F., 40 FC, 10 fixtures per 1000 S.F.	.78	1.53	2.31
0280	3 watts per S.F., 60 FC, 15 fixtures per 1000 S.F	1.17	2.32	3.49
0320	4 watts per S.F., 80 FC, 20 fixtures per 1000 S.F.	1.56	3.08	4.64
0400	5 watts per S.F., 100 FC, 25 fixtures per 1000 S.F.	1.95	3.86	5.81

347

Figure 26.8

Means Forms

COST ANALYSIS

PROJECT: Office Renovation - 3020 S.F.

ESTIMATE NO. 87-2

ARCHITECT: CWL

DATE: 1987

TAKE OFF BY: EBW   QUANTITIES BY: EBW   PRICES BY: RSM   EXTENSIONS BY: SLP   CHECKED BY: JDM

		$/SF Floor	Total Incl. O&P
Summary			
Partitions		2 76	8322
Doors		1 39	4212
Glass		17	515
Wall Finishes		1 86	5621
Floor Finishes		2 73	8254
Ceiling		1 95	5889
HVAC		6 41	19347
Electrical		5 63	17003
Special		1 15	3481
Subtotal		24 05	72644
Contingency	10%	2 41	7264
Total $/SF		26 46	
Total Cost			79908

Figure 26.9

		1.1 Overhead	CREW	DAILY OUTPUT	MAN-HOURS	UNIT	MAT.	LABOR	EQUIP.	TOTAL	TOTAL INCL O&P	
020	0011	ARCHITECTURAL FEES (10)										020
	0020	For work to $10,000				Project					15%	
	0040	To $25,000										
	0060	To $100,000				Project					10%	
	0080	To $500,000									10%	
	0090	To $1,000,000				↓					7%	
040	0010	CLEANING UP After job completion, allow				Job					.30%	040
	0031	Rubbish removal, see division 2.1-430										
	0050	Cleanup of floor area, continuous, per day	A-5	12	1.500	M.S.F.	1.50	24	1.20	26.70	39	
	0100	Final	"	11.50	1.570	"	1.60	25	1.25	27.85	40	
060	0011	CONSTRUCTION COST INDEX For 162 major U.S. and										060
	0020	Canadian cities, total cost, min. (Greensboro, NC)				%					80.40%	
	0050	Average									100%	
	0100	Maximum (Anchorage, AK)				↓					132.60%	
090	0010	CONSTRUCTION MANAGEMENT FEES $1,000,000 job, minimum				Project					4.50%	090
	0050	Maximum									7.50%	
110	0010	CONTINGENCIES Allowance to add at conceptual stage									15%	110
	0050	Schematic stage									10%	
	0100	Preliminary working drawing stage									7%	
	0150	Final working drawing stage				↓					2%	
120	0014	CONTRACTOR EQUIPMENT See division 1.5										120
140	0010	CREWS For building construction, see foreword										140
150	0010	ENGINEERING FEES Educational planning consultant, minimum				Project					.50%	150
	0100	Maximum				"					2.50%	
	0200	Electrical, minimum				Contrct					4.10%	
	0300	Maximum									10.10%	
	0600	Food service & kitchen equipment, minimum									8%	
	0700	Maximum									12%	
	1000	Mechanical (plumbing & HVAC), minimum				↓					4.10%	
	1100	Maximum									10.10%	
180	0010	INSURANCE Builders risk, standard, minimum (2)				Job					.19%	180
	0050	Maximum									1.14%	
	0200	All-risk type, minimum									.20%	
	0250	Maximum									1.16%	
	0600	Public liability, average				↓					.82%	
	0610											
	0800	Workers' compensation & employer's liability, average (7)										
	0850	by trade, carpentry, general				Payroll		11.41%				
	0900	Clerical						.42%				
	0950	Concrete						10.30%				
	1000	Electrical						4.46%				
	1050	Excavation						7.81%				
	1100	Glazing						8.73%				
	1150	Insulation						8.80%				
	1200	Lathing						7.21%				
	1250	Masonry						8.53%				
	1300	Painting & decorating						8.91%				
	1350	Pile driving						18.33%				
	1400	Plastering						8.73%				
	1450	Plumbing						5.31%				
	1500	Roofing						20.77%				
	1550	Sheet metal work (HVAC)						7%				
	1600	Steel erection, structural						21.51%				
	1650	Tile work, interior ceramic						5.96%				
	1700	Waterproofing, brush or hand caulking						5.09%				
	1800	Wrecking						23.08%				
	2000	Range of 36 trades in 50 states, excl. wrecking, minimum				↓		1.10%				
	2100	Average						10.42%				

For expanded coverage of these items see *Means Building Construction Cost Data 1987*

1

Figure 26.10

INDEX